Lecture Notes in Mathematics 1549

Gennadi Vainikko

Multidimensional Weakly Singular Integral Equations

Springer-Verlag
Berlin Heidelberg New York
London Paris Tokyo
Hong Kong Barcelona
Budapest

Author

Gennadi Vainikko
Department of Mathematics
University of Tartu
EE 2400 Tartu, Estonia

Mathematics Subject Classification (1991): 45-02, 45M05, 45L10, 65R20, 35Q60

ISBN 3-540-56878-6 Springer-Verlag Berlin Heidelberg New York
ISBN 0-387-56878-6 Springer-Verlag New York Berlin Heidelberg

Library of Congress Cataloging-in-Publication Data

Vainikko, G.
Multidimensional weakly singular integral equations/Gennadi Vainikko. p. cm. -
(Lecture notes in mathematics; 1549)
Includes bibliographical references and index.
ISBN 3-540-56878-6 (Berlin: acid-free): - ISBN 0-387-56878-6
(New York: acid-free)
1. integral equations-Asymptotic theory. I. Title. II. Series: Lecture notes in
mathematics (Springer-Verlag); 1549.
QA3.L28 no. 1549 (QA431) 510 S-dc20 (515'.45) 93-14009 CIP

46/3140-543210 - Printed on acid-free paper

In memory
of Solomon Mikhlin

PREFACE

In these lecture notes we deal with the integral equation

$$u(x) = \int_G K(x,y)\,u(y)dy + f(x), \quad x \in G, \tag{0.1}$$

where $G \subset \mathbb{R}^n$ is an open bounded region or, more generally, an open bounded set (possibly non-connected). The functions f and K are assumed to be smooth but K may have a weak singularity on the diagonal:

$$|K(x,y)| \le b(1 + |x-y|^{-\nu}), \qquad b = \text{const}, \quad \nu < n. \tag{0.2}$$

The main problems of interest to us are the following:

— the smoothness of the exact solution to equation (0.1);

— discretization methods for equation (0.1).

Usually, the derivatives of the solution to a weakly singular integral equation have singularities near the boundary ∂G of the domain of integration $G \subset \mathbb{R}^n$. A unified description of the singularities in all possible cases is complicated, and up to now this problem has not been solved fully. In Chapter 3 we give estimates which are sharp in many practically interesting cases. The behaviour of the tangential derivatives thereby turns out to be less singular than the behaviour of the normal derivatives. All this information is used designing approximate methods for integral equation (0.1). We restrict ourselves to collocation and related schemes, thoroughly examining simplest schemes based on the piecewise constant approximation of the solution and the superconvergence phenomenon at the collocation points (Chapters 5 and 6). In the case where $G \subset \mathbb{R}^n$ is a parallelepiped, higher order collocation methods on graded grids are also considered (Chapter 7); again the superconvergence at the collocation points is examined.

Technically, our convergence analysis is based on the discrete convergence theory outlined in Chapter 4 of the book. This short chapter can be used for a first acquaintance with the theory for linear equations $u = Tu + f$; for eigenvalue problems and nonlinear equations, the results are presented without proofs.

In Chapter 8, some of the main results of Chapters 3 and 5–7 are extended to nonlinear integral equations.

Examples of (linear) integral equations (0.1), (0.2) can be found in radiation transfer theory (see Section 1); some interior-exterior boundary value problems too have their most natural formulations as integral equations of type (0.1), (0.2). Perhaps some readers will be disappointed to find that our treatment concerns only integral equations on an open set $G \subset \mathbb{R}^n$. In practice, there is a great interest also in the boundary integral equations

$$u(x) = \int_{\partial G} K(x,y)\, u(y)\, dS_y + f(x), \qquad x \in \partial G. \tag{0.3}$$

Such equations arise, for instance, in solving the Dirichlet or Neumann problem for the Laplace equation (see e.g. Mikhlin (1970) or Atkinson (1990)). A natural question of whether the results of the lecture book can be extended or modified to boundary integral equations then arises. The answer is non-unique. If ∂G is smooth then the solution of the boundary integral equation is smooth too, and the results concerning the collocation and related methods can even be strengthened and the arguments can be simplified. On the other hand, if ∂G is non-smooth then the standard boundary integral operators, e.g. the ones corresponding to the Laplace equation, are non-compact, and our arguments fail fully. The case of an integral equation on a smooth (relative) region $\Gamma \subset \partial G$ with a smooth (relative) boundary $\partial\Gamma$ seems to be the most adequate case that can be treated by our arguments. But this assertion may be considered only as a conjecture not discussed anywhere.

We use only a minimum of references in the main text. Nevertheless, an extended commented bibliography is added. Young mathematicians looking for problems to work on will find a list of unsolved problems too. The lectures are based on the author's recent publications (see Vainikko (1990a,b), (1991a,b), (1992a,b), Vainikko and Pedas (1990)) but actually the results were elaborated during a much longer time lecturing at University of Tartu, the Technical University of Chemnitz and Colorado State University. A significant milestone for us was the booklet by Vainikko, Pedas and Uba (1984) concerning the one-dimensional case ($n=1$). In the present lectures, we always assume that $n \geq 2$.

CONTENTS

1. SOME PROBLEMS LEADING TO MULTIDIMENSIONAL WEAKLY SINGULAR INTEGRAL EQUATIONS

In this chapter we present two examples on problems of mathematical physics which can be reformulated as multidimensional weakly singular integral equations — an interior-exterior boundary value problem and a radiation transfer problem.

1.1. An interior-exterior problem. Let $G \subset \mathbb{R}^n$, $n \geq 2$, be an open bounded set with piecewise smooth boundary ∂G and let a and f be given real or complex valued bounded continuous functions on G (we write $a, f \in BC(G)$). Consider the following problem: find a function $\varphi \in C^1(\mathbb{R}^n) \cap H^2_{loc}(\mathbb{R}^n)$ such that

$$\Delta\varphi(x) = a(x)\varphi(x) + f(x)\,, \quad x \in G, \tag{1.1}$$

$$\Delta\varphi(x) = 0, \quad x \in \mathbb{R}^n \setminus \overline{G} \tag{1.2}$$

whereby, in case $n \geq 3$,

$$\varphi(x) \to 0 \quad \text{as} \quad |x| \to \infty \tag{1.3}$$

or, in case $n = 2$,

$$|\varphi(x)| \text{ is bounded as } |x| \to \infty. \tag{1.3'}$$

Here the following standard notations are adopted: $C^1(\mathbb{R}^n)$ is the space of continuously differentiable functions on $\mathbb{R}^n$; $H^2_{loc}(\mathbb{R}^n)$ is the space of functions on $\mathbb{R}^n$ which have locally square-integrable (generalized) derivatives up to the second order; Δ is the Laplace operator, $\Delta\varphi = \partial^2\varphi/\partial x_1^2 + \ldots + \partial^2\varphi/\partial x_n^2$. Note that the condition $\varphi \in C^1(\mathbb{R}^n)$ contains a requirement that φ itself as well its first normal derivative have equal boundary values as x approaches ∂G from inside and outside of G.

1.2. A physical background (n = 2). Crouseix and Descloux (1988) describe a mathematical model of the electromagnetic casting process. When the ingot is sufficiently long, the electromagnetic part of the problem reduces to the search of a complex potential φ in $\mathbb{R}^2$, of class C^1, satisfying the conditions

$$\Delta\varphi + 2i\alpha^2(\varphi + c_k) = 0 \quad \text{in } G_k \ (k = 1, \ldots, l),$$

$$\Delta\varphi = 0 \quad \text{in } \mathbb{R}^2 \setminus \overline{G}, \quad G = \bigcup_{k=1}^{l} G_k.$$

Here $G_k \subset \mathbb{R}^2$, $1 \le k \le l$, are the cross-sections in the x_1, x_2 plane of cylindrical electric conductors in which a current with angular freguency ω runs; $2\alpha^2 = \mu_0 s\omega$ is a real constant where μ_0 is the magnetic permeability of the air and s the conductivity; $i = \sqrt{-1}$ is the imaginary unit and c_k's are given complex constants. Thus, we have a special case of problem (1.1), (1.2) with $a(x) = -2i\alpha^2$, $f(x) = -2i\alpha^2 c_k$ for $x \in G_k$ $(k = 1, \dots, l)$.

An instruction from this background is that we ought to avoid an assumption about the connectivity of $G \subset \mathbb{R}^n$ when problem (1.1), (1.2) will be discussed.

1.3. Integral equation formulation ($n \ge 3$). We look for a solution of problem (1.1)-(1.3) in the form of the Newton potential (see Bers et al. (1964))

$$\varphi(x) = -c_n \int_G |x-y|^{-(n-2)} u(y)\,dy, \qquad x \in \mathbb{R}^n, \tag{1.4}$$

where $c_n = 1/((n-2)\sigma_n)$, $\sigma_n = \Gamma(n/2)/(2\pi^{n/2})$ is the area of the unit sphere in $\mathbb{R}^n$ and $u \in BC(G)$ is the density which we have to determine. Condition (1.3) is automatically fulfilled. First derivatives of φ can be found differentiating (1.4) under the integral sign, the result is a weakly singular integral again, and it is easy to see that $\varphi \in C^1(\mathbb{R}^n)$. Further, it is well known that

$$\Delta\varphi(x) = \begin{cases} u(x), & x \in G, \\ 0, & x \in \mathbb{R}^n \setminus \overline{G} \end{cases}$$

(in the sense of distributions as well in the sense of pointwise equalities). We see that (1.2) is fulfilled, too. A consequence of $\Delta\varphi \in L^2(\mathbb{R}^n)$ is that $\varphi \in L^2_{loc}(\mathbb{R}^n)$ (together with (1.3) we have even $\varphi \in H^2(\mathbb{R}^n)$). Condition (1.1) takes the form of the following integral equation to determine u:

$$u(x) = -c_n a(x) \int_G |x-y|^{-(n-2)} u(y)\,dy + f(x), \quad x \in G. \tag{1.5}$$

Thus, to solve problem (1.1)-(1.3), we have to solve integral equation (1.5) and then apply formula (1.4). Actually, (1.4) is needed only for $x \in \mathbb{R}^n \setminus \overline{G}$; for $x \in G$ we have from (1.4) and (1.5)

$$\varphi(x) = (u(x) - f(x))/a(x).$$

Let us make sure that we exhaust all solutions of problem (1.1)-(1.3) in this way. Indeed, let $\varphi \in C^1(\mathbb{R}^n) \cap H^2_{loc}(\mathbb{R}^n)$ be an arbitrary solution of (1.1)-(1.3). Denote

$$u(x) = \Delta\varphi(x) = a(x)\varphi(x) + f(x), \quad x \in G,$$

$$\psi(x) = \varphi(x) + c_n \int_G |x-y|^{-(n-2)} u(y)\,dy, \quad x \in \mathbb{R}^n.$$

It is clear that $u \in BC(G)$, $\psi \in C^1(\mathbb{R}^n) \cap H^2_{loc}(\mathbb{R}^n)$. In addition , $\Delta\psi(x)=0$ for $x \in G$ and $x \in \mathbb{R}^n \setminus \overline{G}$, i.e. $\Delta\psi=0$ a.e. in $\mathbb{R}^n$. Together with the smoothness of ψ (see above), this means that $\Delta\psi=0$ in $\mathbb{R}^n$ in the sense of distributions. Now, using the hypoellipticity property of the Laplace operator (see e.g. Yosida (1965) or Lions and Magenes (1968)), we obtain that $\psi \in C^\infty(\mathbb{R}^n)$ and $\Delta\psi(x)=0$ for all $x \in \mathbb{R}^n$. Further, $\psi(x) \to 0$ as $|x| \to \infty$, hence $\psi(x)=0$ for all $x \in \mathbb{R}^n$, i.e. φ has a representation (1.4) with $u(x)=\Delta\varphi(x)$, $x \in G$, q.e.d.

Problem (1.1) - (1.3) is uniquely solvable if and only if integral equation (1.5) has a unique solution $u \in BC(G)$. This occurs if and only if the corresponding homogeneous integral equation $u = Tu$ has in $BC(G)$ only the trivial solution. Note that operator $T: BC(G) \to BC(G)$ is compact (a proof in a more general setting is given in Section 2.3)

1.4. Integral equation formulation (n=2). We look for a solution of problem (1.1), (1.2), (1.3') in the form

$$\varphi(x) = (2\pi)^{-1} \int_G \log|x-y| u(y)dy + \beta, \qquad x \in \mathbb{R}^2, \tag{1.6}$$

where we have to determine the density $u \in BC(G)$ and the constant β. Again, $\varphi \in C^1(\mathbb{R}^2)$, $\Delta\varphi(x) = u(x)$ for $x \in G$, $\Delta\varphi(x)=0$ for $x \in \mathbb{R}^2 \setminus \overline{G}$, (1.2) is fulfilled and (1.1) takes the form

$$u(x) = (2\pi)^{-1} a(x) \int_G \log|x-y| u(y)dy + \beta a(x) + f(x), \qquad x \in G. \tag{1.7}$$

Condition (1.3') is fulfilled if and only if

$$\int_G u(x)dx = 0. \tag{1.8}$$

Indeed, rewrite (1.6) in the form

$$\varphi(x) = (2\pi)^{-1} \int_G \log\frac{|x-y|}{|x|} u(y)dy + (2\pi)^{-1} \log|x| \int_G u(y)dy + \beta.$$

Here the first integral tends to 0 as $|x| \to \infty$ since $\log(|x-y|/|x|) \to 0$ uniformly with respect to $y \in G$. Hence $\varphi(x)$ is bounded as $|x| \to \infty$ if and only if (1.8) holds.

Thus, to solve (1.1), (1.2), (1.3'), we have to find a pair $u \in BC(G)$, $\beta \in \mathbb{C}$ (or $\mathbb{R}$) from equations (1.7), (1.8) and then apply (1.6). For $x \in G$ we have $\varphi(x) = (u(x) - f(x))/a(x)$ again, thus actually (1.6) is needed for $x \in \mathbb{R}^2 \setminus \overline{G}$ only. It is easy to check again that we exhaust in this way all solutions of problem (1.1), (1.2), (1.3').

Problem (1.1), (1.2), (1.3') is uniquely solvable if and only if problem (1.7), (1.8) is uniquely solvable. Problem (1.7), (1.8) preserves the Fredholm property — for its unique solvability, it is necessary and sufficient that the corresponding homogeneous problem

$$u(x) = (2\pi)^{-1}a(x)\int_G \log|x-y|\,u(y)dy + \beta a(x), \quad x\in G,$$

$$\int_G u(x)dx = 0$$

has in $BC(G)\times\mathbb{C}$ only the trivial solution $u=0$, $\beta=0$. We state a simple sufficient condition for the unique solvability:

$$a, 1/a\in BC(G), \quad \operatorname{Im} a \neq 0 \text{ and is sign constant in } G \tag{1.9}$$

where $\operatorname{Im} a$ is the imaginary part of a. Indeed, let $u\in BC(G)$, $\beta\in\mathbb{C}$ be a solution of the homogeneous problem. Then we have the equalities

$$\frac{u(x)}{|a(x)|^2}\,\bar{a}(x) = \frac{1}{2\pi}\int_G \log|x-y|\,u(y)dy + \beta, \quad x\in G,$$

$$\int_G u(y)dy = 0$$

where $\bar{a}$ is the complex conjugate to a. Taking the scalar product of the first equality with u and using the second one we obtain

$$\int_G \frac{|u(x)|^2}{|a(x)|^2}\,\bar{a}(x)dx = (\Lambda u, u) \tag{1.10}$$

where the number

$$(\Lambda u, u) = (2\pi)^{-1}\int_G\int_G \log|x-y|\,u(y)\,\bar{u}(x)\,dy\,dx$$

is real due to the symmetry of the kernel $\log|x-y|$. Since $\operatorname{Im} a(x) > 0$ or $\operatorname{Im} a(x) < 0$ on G, (1.10) is possible in case $u=0$ only. Now we see that $\beta=0$, too, q.e.d.

Note that for the physical problem considered in Section 1.2, condition (1.9) is fulfilled.

A further sufficient condition for the unique solvability of the problem (1.7), (1.8) can be formulated as follows: (i) the homogeneous integral equation

$$u(x) = (2\pi)^{-1}a(x)\int_G \log|x-y|\,u(y)dy, \quad x\in G,$$

has only the trivial solution $u=0$; (ii) the solution u_a of the integral equation

$$u(x) = (2\pi)^{-1}a(x)\int_G \log|x-y|\,u(y)dy + a(x), \quad x\in G,$$

satisfies the condition $\int_G u_a(x)dx \neq 0$.

If conditions (i) and (ii) are fulfilled then the unique solution of problem (1.7), (1.8) is given by

$$\beta = -\int_G u_f(x)dx \Big/ \int_G u_a(x)dx\,, \quad u(x) = u_f(x) + \beta u_a(x)$$

where u_f is the solution of the integral equation

$$u(x) = (2\pi)^{-1}a(x)\int_G \log|x-y|\,u(y)dy + f(x), \quad x \in G.$$

Thus, problem (1.7), (1.8) can be reduced to two standard integral equations with the logarithmic kernel. But numerically one usually prefers to solve (1.7), (1.8) directly.

1.5. Radiation transfer problem. Let $G \subset \mathbb{R}^3$ be an open bounded convex region, ∂G its boundary and

$$S = \{s \in \mathbb{R}^3 : |s| = (s_1^2 + s_2^2 + s_3^2)^{1/2} = 1\}$$

the unit sphere in $\mathbb{R}^3$. For $x \in \partial G$, let us denote by

$$S_x = \{s \in S : \exists \lambda > 0 : x + \lambda s \in G\}$$

the set of directions falling into G; if x is a smoothness point of ∂G then we simply have

$$S_x = \{s \in S : s \cdot \nu(x) > 0\}$$

where $\nu(x)$ is the unit inner normal to ∂G at $x \in \partial G$.

A standard radiation transfer problem reads as follows: find a function $\varphi : \overline{G} \times S \to \mathbb{R}_+$ (the intensity of the radiation) such that

$$\sum_{j=1}^{3} s_j \frac{\partial \varphi(x,s)}{\partial x_j} + \sigma(x)\varphi(x,s) = \frac{\sigma_0(x)}{4\pi}\int_S g(x,s,s')\varphi(x,s')ds' + f(x,s) \quad (x \in G,\, s \in S), \tag{1.11}$$

$$\varphi(x,s) = \varphi_0(x,s) \quad (x \in \partial G,\, s \in S_x). \tag{1.12}$$

Here $\sigma : G \to \mathbb{R}_+$ (the extinction coefficient), $\sigma_0 : G \to \mathbb{R}_+$ (the scattering coefficient), $g : G \times S \times S \to \mathbb{R}_+$ (the phase function of scattering), $f : G \times S \to \mathbb{R}_+$ (the source function) and $\varphi_0 : \{(x,s) : x \in \partial G,\, s \in S_x\} \to \mathbb{R}_+$ (the inflow radiation intensity) are given functions whereby

$$\sigma_0(x) \le \sigma(x), \qquad g(x,s,s') = g(x,s',s), \qquad \frac{1}{4\pi}\int_S g(x,s,s')ds' = 1$$

(we shall refer to those as "physical conditions"). For a more detailed exposition we refer to Chandrasekhar (1950). We have adopted a terminology used in the atmospheric optics. In the theory of nuclear reactors, (1.11),

(1.12) occurs as a main problem, too, but the terminology is slightly different (see e.g. Case and Zweifel (1967) or Marchuk and Lebedev (1984)).

1.6. Integral equation formulation of the radiation transfer problem. Let us split (1.11) into two equations:

$$\sum_{j=1}^{3} s_j \frac{\partial \varphi(x,s)}{\partial x_j} + \sigma(x)\varphi(x,s) = u(x,s) \qquad (x \in G,\ s \in S), \tag{1.11'}$$

$$u(x,s) = \frac{\sigma_0(x)}{4\pi} \int_S g(x,s,s')\varphi(x,s')ds' + f(x,s) \qquad (x \in G,\ s \in S). \tag{1.11''}$$

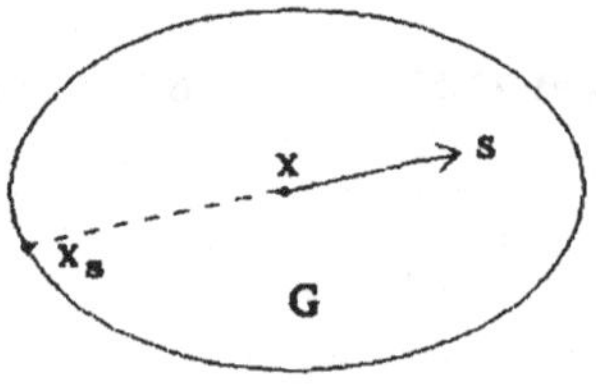

Fig. 1.1

One can solve (1.11'), (1.12) explicitly:

$$\varphi(x,s) = \varphi_0(x_s,s)\exp(-\tau(x,x_s))$$
$$+ \int_{-|x-x_s|}^{0} u(x+\lambda s, s)\exp(-\tau(x, x+\lambda s))d\lambda \qquad (x \in G,\ s \in S) \tag{1.13}$$

where x_s is the point on ∂G which lies in the direction $-s$ from x (see Figure 1.1) and

$$\tau(x,y) = |x-y| \int_0^1 \sigma(tx + (1-t)y)dt \tag{1.14}$$

is the optical distance between points $x,y \in G$.

Indeed, let us write down that φ satisfies (1.11') at the points of the straight line $x+\lambda s$ $(-|x-x_s| < \lambda < 0)$:

$$\sum_{j=1}^{3} s_j \frac{\partial \varphi(x+\lambda s, s)}{\partial x_j} + \sigma(x+\lambda s)\varphi(x+\lambda s, s) = u(x+\lambda s, s).$$

Since

$$\sum_{j=1}^{3} s_j \frac{\partial \varphi(x+\lambda s, s)}{\partial x_j} = \frac{d}{d\lambda}\varphi(x+\lambda s, s),$$

we obtain the linear ordinary differential equation of the first order

$$\frac{d}{d\lambda}\varphi(x+\lambda s, s) + \sigma(x+\lambda s)\varphi(x+\lambda s, s) = u(x+\lambda s, s), \quad -|x-x_s| \le \lambda \le 0,$$

with respect to $\varphi(x+\lambda s,s)$ as a function of λ. Condition (1.12) yields an initial data:

$$\varphi(x+\lambda s,s) = \varphi_0(x_s,s) \quad \text{for} \quad \lambda = -|x-x_s|.$$

Solving this Cauchy problem and putting $\lambda = 0$ in the solution we get formula (1.13).

Substituting (1.13) into (1.11") we obtain

$$u(x,s) = \frac{\sigma_0(x)}{4\pi} \int_S g(x,s,s') \int_{-|x-x_{s'}|}^{0} u(x+\lambda s',s') \exp(-\tau(x,x+\lambda s')) d\lambda ds'$$

$$+ f(x,s) + \frac{\sigma_0(x)}{4\pi} \int_S g(x,s,s') \varphi_0(x_{s'},s') \exp(-\tau(x,x_{s'})) ds'.$$

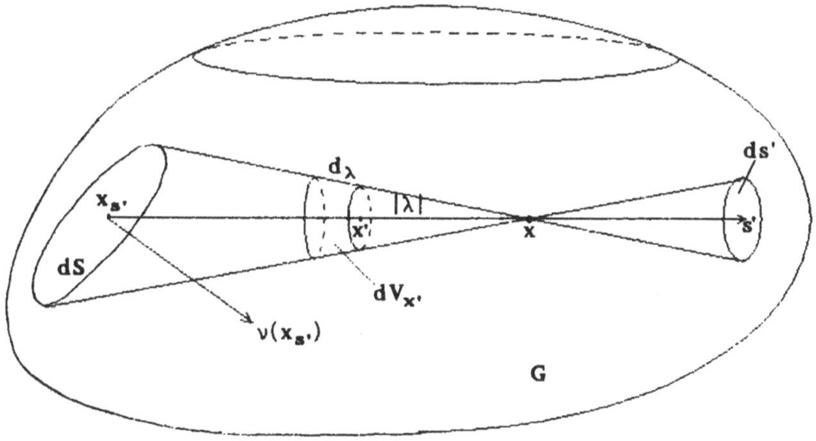

Fig. 1.2.

Denoting $x' = x+\lambda s' \in G$ $(-|x-x_{s'}| < \lambda < 0)$ we have $s' = (x-x')/|x-x'|$; taking into account also that (see Figure 1.2)

$$|x-x'|^2 ds' d\lambda = dV_{x'} = dx' \quad \text{(the volume element in G)},$$

$$|x-x_{s'}|^2 \frac{ds'}{s' \cdot \nu(x_{s'})} = dS \quad \text{(the area element on } \partial G\text{)},$$

we obtain the integral equation

$$u(x,s) = \frac{\sigma_0(x)}{4\pi} \int_G g\left(x,s,\frac{x-x'}{|x-x'|}\right) \exp(-\tau(x,x')) |x-x'|^{-2} u\left(x',\frac{x-x'}{|x-x'|}\right) dx'$$
$$+ f(x,s) + f_0(x,s) \qquad (x \in G,\ s \in S) \tag{1.15}$$

where

$$f_0(x,s) = \frac{\sigma_0(x)}{4\pi} \int_{\partial G} g(x,s,\frac{x-x'}{|x-x'|}) \exp(-\tau(x,x'))\,|x-x'|^{-2}$$

$$\times \frac{\nu(x')\cdot(x-x')}{|x-x'|} \varphi_0(x',\frac{x-x'}{|x-x'|})\,dS_{x'} \qquad (x\in G,\ s\in S). \tag{1.16}$$

Finding $u: G\times S \to \mathbb{R}$ from (1.15) one obtains the solution φ of problem (1.11), (1.12) via formula (1.13). Under physical conditions on σ, σ_0 and g formulated in Section 1.5, together with the boundness of those functions, one can show that the integral operator $T\in\mathcal{L}(L^\infty(G\times S), L^\infty(G\times S))$ of equation (1.15) has a spectral radius $\rho(T)<1$.

1.7. Peierls integral equation. Consider a special case where $g(x,s,s')=1$ for all $x\in G$, $s,s'\in S$ (the isotropic scattering) and f does not depend on s. Then (1.15) reduces to the integral equation

$$u(x) = \frac{\sigma_0(x)}{4\pi} \int_G \exp(-\tau(x,y))\,|x-y|^{-2} u(y)\,dy + f(x) + f_0(x), \qquad x\in G, \tag{1.17}$$

which is known as Peierls integral equation. Here

$$f_0(x) = \frac{\sigma_0(x)}{4\pi} \int_{\partial G} \exp(-\tau(x,y))\,|x-y|^{-2} \frac{\nu(y)\cdot(x-y)}{|x-y|} \varphi_0(y,\frac{x-y}{|x-y|})\,dS_y, \quad x\in G. \tag{1.18}$$

1.8. Exercises. 1.8.1. Find the integral equation formulations of the interior-exterior problem where (1.1) is replaced by a more general equation

$$\Delta\varphi(x) = \sum_{j=1}^{n} a_j(x) \frac{\partial\varphi(x)}{\partial x_j} + a(x)\varphi(x) + f(x), \qquad x\in G.$$

1.8.2. Prove that problem (1.7), (1.8) is uniquely solvable in the following singular case, too: the homogeneous integral equation $u=Tu$ has an one dimensional solution space but $\int_G u(x)dx \neq 0$ for a solution. If the solution space is two or more dimensional then problem (1.7), (1.8) is nonsolvable or has more than one solution.

1.8.3. We succeeded in integral equation reformulations in Sections 1.3, 1.4 knowing the fundamental solution F(x,y) of the Laplace operator ($F(x,y) = -c_n|x-y|^{-n+2}$ for $n\geq 3$, $F(x,y) = (2\pi)^{-1}\log|x-y|$ for $n=2$). Can you present other examples of differential operators with known fundamental solutions? Formulate corresponding interior-exterior problems and their integral equation counterparts.

1.8.4. Let φ be the solution of problem (1.11), (1.12). Show that

$$v(x,s) = \frac{1}{4\pi} \int_S g(x,s,s')\,\varphi(x,s')\,ds'$$

satisfies the integral equation

$$v(x,s) = \frac{1}{4\pi}\int_S \sigma_0(x')\, g(x,s,\frac{x-x'}{|x-x'|})\, \exp(-\tau(x,x'))\,|x-x'|^{-2}\, v(x',\frac{x-x'}{|x-x'|})\,dx'$$

$$+\ \tilde{f}(x,s) + \tilde{f}_0(x,s) \qquad (x\in G,\ s\in S) \tag{1.15'}$$

where

$$\tilde{f}(x,s) = \frac{1}{4\pi}\int_G g(x,s,\frac{x-x'}{|x-x'|})\, \exp(-\tau(x,x'))\,|x-x'|^{-2}\, f(x',\frac{x-x'}{|x-x'|})\,dx',$$

$$\tilde{f}_0(x,s) = \frac{1}{4\pi}\int_{\partial G} g(x,s,\frac{x-x'}{|x-x'|})\exp(-\tau(x,x'))\,|x-x'|^{-2}\,\frac{\nu(x')\cdot(x-x')}{|x-x'|}\,\varphi_0(x',\frac{x-x'}{|x-x'|})\,dS_{x'}.$$

In the case of the isotropic scattering, instead of (1.17) one obtains the Peierls integral equation

$$v(x) = \frac{1}{4\pi}\int_G \sigma_0(y)\exp(-\tau(x,y))\,|x-y|^{-2}\,v(y)\,dy + \tilde{f}(x) + \tilde{f}_0(x) \tag{1.17'}$$

where

$$\tilde{f}(x) = \frac{1}{4\pi}\int_G \exp(-\tau(x,y))\,|x-y|^{-2}\, f(y,\frac{x-y}{|x-y|})\,dy,$$

$$\tilde{f}_0(x) = \frac{1}{4\pi}\int_{\partial G} \exp(-\tau(x,y))\,|x-y|^{-2}\,\frac{\nu(y)\cdot(x-y)}{|x-y|}\,\varphi_0(y,\frac{x-y}{|x-y|})\,dS_y .$$

1.8.5. Prove that $\|T'\|<1$ where $T'\in\mathcal{L}(L^\infty(G\times S),L^\infty(G\times S))$ is the integral operator of equation (1.15'). Thereby one may assume that σ, σ_0 and g are bounded and satisfy the physical conditions formulated in Section 1.5. Imply from here that $\rho(T)<1$, the assertion in the end of Section 1.6.

2. PRELIMINARIES

In this Chapter we introduce some function spaces which will be used during the whole book. Some general results concerning weakly singular integral operators are presented, too.

2.1. Metric d_G and space $BC(G^*)$. Let $G \subset \mathbb{R}^n$ be an open bounded set. Being open, G consists from a finite or countable number of connectivity components (maximal connected open subsets of G). For $x,y \in G$ belonging to a common connectivity component of G, define the "inner" distance $d_G(x,y)$ as the infimum of the lengths of the polygonal paths in G joining these points (see Figure 2.1). For $x,y \in G$ belonging to different connectivity components of G, put $d_G(x,y) = d_*$ where d_* is the supremum of $d_G(x^1,x^2)$ as x^1 and x^2 vary in common connectivity components. Note that $d_* = \infty$ is possible although $G \subset \mathbb{R}^n$ is bounded; for instantce, $d_* = \infty$ for G presented on Figure 2.1.

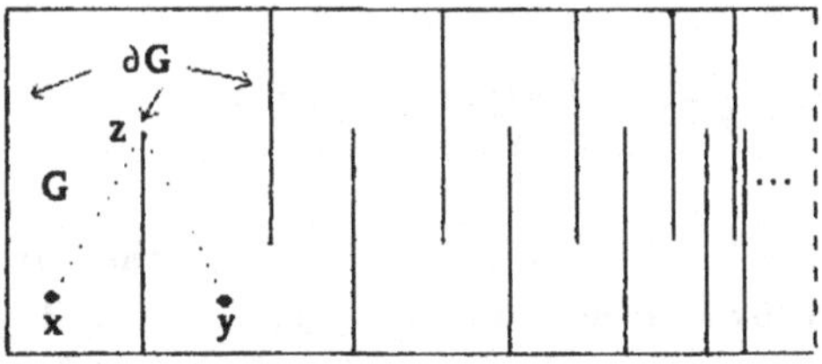

Fig. 2.1. $d_G(x,y) = |x-z| + |z-y|$

As usually, we denote by $\overline{G}$ the closure and by $\partial G = \overline{G} \setminus G$ the boundary of G with respect to the standard Euclidean norm in $\mathbb{R}^n$, $|x| = (x_1^2 + \ldots + x_n^2)^{1/2}$. Let us denote by G^* the completion of G in the metric d_G. In the case of not too complicated structure of G, the d_G-boundary $\partial^* G = G^* \setminus G$ can be interpreted geometrically as "multiple boundary ∂G ". Namely, let $G_1(x^0,\varepsilon), \ldots, G_k(x^0,\varepsilon)$ be the connectivity components of the set $G \cap B(x^0,\varepsilon)$, where $x^0 \in \partial G$ and $\varepsilon = \varepsilon(x^0) > 0$ is small enough that making it smaller does not change the number $k = k(x^0)$ of connectivity components. A sequence $x^{k,j} \in G_j(x^0,\varepsilon)$ with $|x^{k,j} - x^0| \to 0$ as $k \to \infty$ is Cauchy in the metric d_G; we assign to it a limit $x^{0,j} \in G^*$, "a copy of the point x^0". Thus, corresponding to a point $x^0 \in \partial G$ are the points $x^{0,j} \in \partial^* G$, $j = 1, \ldots, k(x^0)$. On Figure 2.1, the inner boundaries occur in $\partial^* G$ twice, the outer boundary partly once whereby a part of ∂G given as a dash line does not occur in $\partial^* G$ at all.

In general, G^* need not be compact in the metric d_G. For instance, if G has a countable number of connectivity components then G^* is non-compact. For a connected region G on Figure 2.1, G^* is non-compact, too. Nevertheless, for problems from practice, G^* is usually compact.

In Section 1.1 we always introduced the space $BC(G)$ of bounded continuous functions on G; let us equip it with the norm

$$\|u\|_{BC(G)} = \sup_{x\in G} |u(x)| = \|u\|_{L^\infty(G)}.$$

In a similar way we introduce the space $BC(G^*)$. It consists of bounded continuous functions on G^*,

$$\|u\|_{BC(G^*)} = \sup_{x\in G^*} |u(x)| = \sup_{x\in G} |u(x)| = \|u\|_{L^\infty(G)}.$$

A function $u \in BC(G^*)$ has boundary values on an inner boundary of G but these values may be different from different sides of the inner boundary. Thus, the space $BC(G^*)$ is useful when piecewise continuous functions on $\overline{G}$ are considered.

We have

$$C(\overline{G}) \subset BC(G^*) \subset BC(G) \subset L^\infty(G) \quad \text{(closed subspaces)}.$$

Note that $BC(G^*) = C(G^*)$ in the case of compact G^*. Moreover, in the case of compact G^*, a function $u \in BC(G)$ can be extended up to a function belonging to $BC(G^*)$ if and only if u is uniformly continuous on G with respect to the metric d_G:

$$\forall \varepsilon > 0 \quad \exists \delta = \delta(\varepsilon) > 0: \quad x^1, x^2 \in G,\ d_G(x^1, x^2) < \delta \quad \Rightarrow \quad |u(x^1) - u(x^2)| < \varepsilon.$$

In the case of non-compact G^*, this condition remains to be sufficient for $u \in BC(G)$ to be extendible up to a function belonging to $BC(G^*)$.

2.2. Weakly singular integral operators. In this book we shall consider the integral equations of the second kind $u = Tu + f$ with integral operators of the type

$$(Tu)(x) = \int_G K(x,y)\,u(y)\,dy$$

where $G \subset \mathbb{R}^n$ is an open bounded set and $K(x,y)$ is a continuous kernel on $(G\times G)\setminus\{x=y\}$ which may have a weak singularity on the diagonal $x = y$:

$$|K(x,y)| \le b\,|x-y|^{-\nu} \qquad (x,y \in G), \quad 0 < \nu < n, \quad b = \text{const}. \tag{2.1}$$

Let us emphasize the condition $\nu < n$ in (2.1) which guarantees the absolute integrability of the kernel (the "weakness" of the singularity). Namely

$$\int_G |K(x,y)|\,dy \le b\sigma_n \frac{d^{n-\nu}}{n-\nu} \qquad (x\in G)$$

where d is the diameter of G and σ_n is the area of the unit sphere in $\mathbb{R}^n$. Indeed, the volume of an infinite thin layer $r<|y|<r+dr$ is $dy=\sigma_n r^{n-1}dr$, therefore

$$\int_G |K(x,y)|\,dy \le b \int_{\{y\in\mathbb{R}^n:\,|x-y|<d\}} |x-y|^{-\nu}dy = b \int_{|y|<d} |y|^{-\nu}dy$$

$$= b\sigma_n \int_0^d r^{-\nu} r^{n-1} dr = b\sigma_n \frac{d^{n-\nu}}{n-\nu}\,.$$

2.3. Compactness of a weakly singular integral operator. Our main consideration will concern a case where the kernel K(x,y) is differentiable and the derivatives satisfy some estimates. Here we confine ourselves to some kind of uniform continuity conditions with respect to the metric d_G:

$$\left.\begin{array}{l}\forall\varepsilon>0\ \ \forall h>0\ \ \exists\delta=\delta(\varepsilon,h)>0:\ x^1,x^2,y\in G,\ |x^i-y|\ge h\ \ (i=1,2),\\ d_G(x^1,x^2)<\delta \ \Rightarrow\ |K(x^1,y)-K(x^2,y)|<\varepsilon;\end{array}\right\} \tag{2.2}$$

$$\left.\begin{array}{l}\forall\varepsilon>0\ \ \forall h>0\ \ \exists\delta=\delta(\varepsilon,h)>0:\ x,y^1,y^2\in G,\ |x-y^i|\ge h\ \ (i=1,2),\\ d_G(y^1,y^2)<\delta \ \Rightarrow\ |K(x,y^1)-K(x,y^2)|<\varepsilon.\end{array}\right\} \tag{2.3}$$

These conditions allow discontinuities of K with respect to x and y on possible inner boundaries of G.

Lemma 2.1. Under conditions (2.1) and (2.2), T is a linear bounded operator from $L^p(G)$, $p>n/(n-\nu)$, to $BC(G^*)$.

Proof. Repeating the argument of Section 2.2 we find that

$$\sup_{x\in G}\Big(\int_G |K(x,y)|^q dy\Big)^{1/q} \le b\Big(\sigma_n \frac{d^{n-q\nu}}{n-q\nu}\Big)^{1/q}, \qquad 1\le q<n/\nu. \tag{2.4}$$

For $p>n/(n-\nu)$, its Hölder conjugate number q ($p^{-1}+q^{-1}=1$) satisfies $q<n/\nu$. Using the Hölder inequality and (2.4) we estimate

$$\Big|\int_G K(x,y)u(y)\,dy\Big| \le b\Big(\sigma_n\frac{d^{n-q\nu}}{n-q\nu}\Big)^{1/q}\|u\|_{L^p(G)}.$$

Thus, T is a linear bounded operator from $L^p(G)$, $p>n/(n-\nu)$, to $L^\infty(G)$ whereby

$$\|T\|_{\mathcal{L}(L^p(G),L^\infty(G))} \le b\Big(\sigma_n\frac{d^{n-q\nu}}{n-q\nu}\Big)^{1/q}, \qquad p^{-1}+q^{-1}=1.$$

To complete the proof of the Lemma, it remains to show that $Tu\in BC(G^*)$

for any $u \in L^p(G)$, $p > n/(n-\nu)$.

The Hölder inequality yields

$$|(Tu)(x^1)-(Tu)(x^2)| = \Big|\int_G [K(x^1,y)-K(x^2,y)]u(y)dy\Big| \le$$

$$\le \Big(\int_G |K(x^1,y)-K(x^2,y)|^q dy\Big)^{1/q} \|u\|_{L^p(G)}.$$

To obtain that $Tu \in BC(G^*)$, it sufficies to show that

$$\sup_{x^1,x^2 \in G, d_G(x^1,x^2)<\delta} \Big(\int_G |K(x^1,y)-K(x^2,y)|^q dy\Big)^{1/q} \to 0 \quad \text{as} \quad \delta \to 0. \tag{2.5}$$

Intersecting G with a ball $B(x^1,h)=\{y \in \mathbb{R}^n: |y-x^1|<h\}$, (2.4) reduces to

$$\sup_{x \in G \cap B(x^1,h)} \Big(\int_{G \cap B(x^1,h)} |K(x,y)|^q dy\Big)^{1/q} \le ch^{(n-q\nu)/q}, \quad 1 \le q < n/\nu,$$

where the constant $c = c_{b,n,\nu,q}$ does not depend on h. Now (2.5) is a consequence of condition (2.2):

$$\Big(\int_G |K(x^1,y)-K(x^2,y)|^q dy\Big)^{1/q} \le$$

$$\Big(\int_{G\setminus(B(x^1,2h)\cap B(x^2,2h))} |K(x^1,y)-K(x^2,y)|^q dy\Big)^{1/q}$$

$$+ \Big(\int_{G\cap B(x^1,2h)} |K(x^1,y)|^q dy\Big)^{1/q} + \Big(\int_{G\cap B(x^2,2h)} |K(x^2,y)|^q dy\Big)^{1/q}$$

$$\le \varepsilon(\operatorname{meas} G)^{1/q} + 2c(2h)^{(n-q\nu)/q}$$

with arbitrary $\varepsilon>0$ and $h>0$ if $d_G(x^1,x^2)<\delta\le h$ (implying $|x^1-x^2|<h$ and $|x^i-y|\ge h$ $(i=1,2)$ for $y \in G\setminus(B(x^1,2h)\cap B(x^2,2h))$ where $\delta=\delta(\varepsilon,h)>0$ is chosen as required in (2.2).

Lemma 2.1 is proved.

It follows from (2.4) and (2.5) that T maps the unit ball of $L^p(G)$ into a set of uniformly bounded d_G-equicontinuous functions. If G^* is compact then, due to Arzelà Lemma, this set is relatively compact in the space $BC(G^*)$. Hence the following corollary holds.

Corollary 2.1. If G^* is compact then, under conditions (2.1) and (2.2), the operator $T \in \mathcal{L}(L^p(G),C(G^*))$, $p>n/(n-\nu)$, is compact.

In the following Lemma, condition (2.3) relieves the compactness condition of G^*.

Lemma 2.2. Under conditions (2.1), (2.2) and (2.3), the operator $T \in \mathcal{L}(L^p(G), BC(G^*))$, $p > n/(n-\nu)$, is compact.

Proof. Let $(u_j) \subset L^p(G)$, $n/(n-\nu) < p < \infty$, $\|u_j\|_{L^p(G)} \le 1$ $(j \in \mathbb{N} = \{1,2,3,\dots\})$; we denote $v_j = Tu_j$. We have to prove that (v_j) contains a subsequence which converges in the $L^\infty(G)$ norm, i.e. uniformly on G.

Since the space $L^p(G)$ with $1 < p < \infty$ is reflexive, (u_j) is weakly compact in $L^p(G)$; let

$$u_j \to u \text{ weakly in } L^p(G) \quad \text{as} \quad j \to \infty,\ j \in N' \subseteq \mathbb{N}.$$

Our purpose is to show that $v_j \to v = Tu$ in the $L^\infty(G)$ norm as $j \to \infty$, $j \in N'$. For any fixed $x \in G$, we have

$$v_j(x) - v(x) = \int_G K(x,y)[u_j(y) - u(y)]dy \to 0 \quad \text{as} \quad j \to \infty,\ j \in N'$$

since $k_x(y) = K(x,y)$ as a function of y belongs to $L^q(G)$, $p^{-1} + q^{-1} = 1$, $q < n/\nu$ (see (2.4)). This convergence is uniform with respect to $x \in G$ if the family $\{k_x(y)\}_{x \in G}$ is relatively compact in $L^q(G)$. In that follows we establish the relative compactness of $\{k_x(y)\}_{x \in G}$.

Denoting by $\rho(x) = \inf_{y \in \partial G} |x-y|$ the Euclidean distance from $x \in G$ to ∂G, we devide G into two subsets

$$\Gamma_h = \{x \in G: \rho(x) < h\}, \quad F_h = \{x \in G: \rho(x) \ge h\}.$$

Due to the continuity property of the Lebesgue measure,

$$\text{meas}\, \Gamma_h \to 0 \quad \text{as} \quad h \to 0.$$

The set F_h is closed in $\mathbb{R}^n$. An important observation is that

$$d_G(x^1, x^2) \le c_h |x^1 - x^2| \quad \text{for any} \quad x^1, x^2 \in F_h \cap G_0 \qquad (2.6)$$

where G_0 is any connectivity component of G and c_h is a constant which may depend on h but not on G_0 and x^1 and x^2 from $F_h \cap G_0$. Indeed, in the opposite case there are $x^j, \bar{x}^j \in F_h \cap G_j$ such that $d_G(x^j, \bar{x}^j) > j|x^j - \bar{x}^j|$, $j = 1,2,\dots$. There is only a finite number of connectivity components G_j of G containing a point x such that $\rho(x) \ge h$, therefore F_h meets only a finite number of those and we may assume that all x^j, $\bar{x}^j$ $(j = 1,2,\dots)$ belong to a common connectivity component G_0 of G. Let $x^j \to x$, $\bar{x}^j \to \bar{x}$ as $j \to \infty$, $j \in N' \subseteq \mathbb{N}$. Then $x, \bar{x} \in G_0$, too, and $d_G(x, \bar{x}) < \infty$. Completing the polygonal paths with the intervals from x to x^j and $\bar{x}$ to $\bar{x}^j$, it is easy to see that

$$d_G(x^j, \bar{x}^j) \le d_G(x, \bar{x}) + |x - x^j| + |\bar{x} - \bar{x}^j| \to d_G(x, \bar{x}) \quad \text{as} \quad j \to \infty,\ j \in N'.$$

Therefore,

$$|x^j - \bar{x}^j| < j^{-1} d_G(x^j, \bar{x}^j) \to 0 \quad \text{as} \quad j \to \infty,\ j \in N'.$$

Hence $x=\bar{x}$ and $d_G(x^j,\bar{x}^j)=|x^j-\bar{x}^j|$ for sufficiently great $j\in N'$. We have got a contradiction which proves (2.6)

From (2.5) and (2.6) we obtain

$$\sup_{x^1,x^2\in F_h\cap G_0,\,|x^1-x^2|<c_h^{-1}\delta} \Big(\int_G |K(x^1,y)-K(x^2,y)|^q dy\Big)^{1/q} \to 0 \quad \text{as} \quad \delta\to 0.$$

Thus, $k_x\in L^q(G)$ depends continuously on $x\in F_h\cap G_0$, and the family $(k_x)_{x\in F_h}$ is compact in $L^q(G)$. Further, for $x\in\Gamma_h$, $y\in F_{2h}$ we have $|x-y|>h$, $|K(x,y)|\le bh^{-\nu}$; condition (2.3) together with (2.6) yields:

$$\forall\varepsilon>0 \quad \forall h>0 \quad \exists\delta=\delta(\varepsilon,h)>0:\ x\in\Gamma_h,\ y^1,y^2\in F_{2h}\cap G_0,\ |y^1-y^2|<c_h^{-1}\delta \ \Rightarrow$$
$$|K(x,y^1)-K(x,y^2)|<\varepsilon.$$

We can use the Arzelà Lemma (with $\mathbb{R}^n$-topology in $F_{2h}\cap G_0$) and we obtain that the family $\{k_x\}_{x\in\Gamma_h}$ is relatively compact in $C(F_{2h})$, and $L^q(F_{2h})$; recall that F_{2h} meets only a finite number of connectivity components of G. In other words, $\{\varphi_h k_x\}_{x\in\Gamma_h}$ is relatively compact in $L^q(G)$ where φ_h is the characteristic function of F_{2h},

$$\varphi_h(y) = \begin{cases} 1, & y\in F_{2h}, \\ 0, & y\in\Gamma_{2h}. \end{cases}$$

Introduce the Hausdorff measure of non-compactness of a set $\mathcal{M}\subset L^q(G)$:

$$\chi(\mathcal{M}) = \inf\{\varepsilon>0:\ \text{there is a relatively compact set } \mathcal{M}_\varepsilon\subset L^q(G) \text{ such that } \sup_{v\in\mathcal{M}}\inf_{v_\varepsilon\in\mathcal{M}_\varepsilon}\|v-v_\varepsilon\|_{L^q(G)}\le\varepsilon\}$$

(see Akhmerov et al. (1992)). Our argument can be summarized as follows: for $\mathcal{M}=\{k_x\}_{x\in G}$ we have a relatively compact set $\mathcal{M}_h=\{k_x\}_{x\in F_h}\cup\{\varphi_h k_x\}_{x\in\Gamma_h}$ such that

$$\sup_{v\in\mathcal{M}}\ \inf_{v_h\in\mathcal{M}_h}\|v-v_h\|_{L^q(G)} = \sup_{x\in\Gamma_h}\|(1-\varphi_h)k_x\|_{L^q(G)}$$

$$= \sup_{x\in\Gamma_h}\Big(\int_{\Gamma_{2h}}|K(x,y)|^q dy\Big)^{1/q} \le b\sup_{x\in\Gamma_h}\Big(\int_{\Gamma_{2h}}|x-y|^{-q\nu}dy\Big)^{1/q}\to 0 \quad \text{as} \quad h\to 0$$

since meas $\Gamma_{2h}\to 0$ as $h\to 0$. This means that $\chi(\mathcal{M})=0$ and $\mathcal{M}=\{k_x\}_{x\in G}$ is relatively compact in $L^q(G)$. The proof of Lemma 2.2 is completed.

Corollary 2.2. Under conditions (2.1) and (2.2), $T\in\mathcal{L}(BC(G),BC(G))$ and $T\in\mathcal{L}(BC(G^*),BC(G^*))$. If (2.3) is fulfilled too then these operators are compact.

Remark 2.1. Without condition (2.3) the operators $T \in \mathcal{L}(BC(G), BC(G))$ and $T \in \mathcal{L}(BC(G^*), BC(G^*))$ need not be compact. Nevertheless, under conditions (2.1) and (2.2) the operators $T^2 \in \mathcal{L}(BC(G), BC(G))$ and $T^2 \in \mathcal{L}(BC(G^*), BC(G^*))$ are compact without condition (2.3).

Let us first present an example where T is non-compact. Define $K(x,y) = \sin(d_G(x,x^0)y_1)$ where $x^0 \in G$ is fixed, $y=(y_1,y_2)$, $n=2$ and $G \subset \mathbb{R}^2$ is the region from Figure 2.1. Having no singularity, $K(x,y)$ satisfies (2.1) with any $\nu > 0$; (2.2) is satisfied too but not (2.3). The family $(k_x)_{x \in G}$ contains the sequence $u_j(y) = \sin jy_1$, $j = 1,2,\ldots$. Since $u_j \to 0$ weakly in $L^2(G)$, also $Tu_j \to 0$ weakly in $L^2(G)$. If $T \in \mathcal{L}(BC(G), BC(G))$ were compact then $Tu_j \to 0$ uniformly in G but this is not the case.

The compactness of T^2 is a corollary of Lemma 2.1 and Remark 2.2.

Remark 2.2. An integral operator with a measurable kernel $K(x,y)$ satisfying (2.1) is compact as an operator from $L^p(G)$ into $L^p(G)$, $1 \le p < \infty$.

We refer to Kantorovich and Akilov (1982) for the proof.

2.4. Weakly singular integral operators with differentiable kernels. Now we assume that the kernel $K(x,y)$ is differentiable on $(G \times G) \setminus \{x=y\}$ and

$$|\partial K(x,y)/\partial x_i| \le b' |x-y|^{-\nu-1}, \quad i=1,\ldots,n, \tag{2.7}$$

$$|\partial K(x,y)/\partial y_i| \le b' |x-y|^{-\nu-1}, \quad i=1,\ldots,n, \tag{2.8}$$

where $\nu \in (0,n)$ is the same number as in (2.1). Our first observation is that (2.7) implies (2.2) whereas (2.8) implies (2.3). Indeed, let $\varepsilon > 0$ and $h > 0$ be given; define $\delta = \min\{h/4, (2n^{1/2}b')^{-1}(h/2)^{\nu+1}\varepsilon\}$. We check that this δ is suitable for (2.2). For any $x^1, x^2, y \in G$ with $|x^i - y| \ge h$ $(i=1,2)$, $d_G(x^1,x^2) < \delta$ we construct a polygonal path $[z^1, z^2, \ldots, z^l] \subset G$ with nodes at points $x^1 = z^1, z^2, \ldots, z^l = x^2$ such that

$$\sum_{k=1}^{l-1} |z^{k+1} - z^k| \le 2d_G(x^1,x^2) \quad (\le h/2).$$

The distance from y to this path remains exceeding h/2. We have

$$\begin{aligned} K(x^2,y) - K(x^1,y) &= \sum_{k=1}^{l-1} [K(z^{k+1},y) - K(z^k,y)] \\ &= \sum_{k=1}^{l-1} \int_0^1 \frac{\partial K(tz^{k+1} + (1-t)z^k, y)}{\partial t}\, dt \qquad (2.9) \\ &= \sum_{k=1}^{l-1} \int_0^1 \sum_{i=1}^{n} \frac{\partial K(tz^{k+1} + (1-t)z^k, y)}{\partial x_i}\, dt\,(z_i^{k+1} - z_i^k), \end{aligned}$$

and with the help of (2.7) we find

$$|K(x^1,y)-K(x^2,y)| \le b'(h/2)^{-\nu-1}\sum_{k=1}^{l-1}\sum_{i=1}^{n}|z_i^{k+1}-z_i^k|$$

$$\le b'(h/2)^{-\nu-1}n^{1/2}\sum_{k=1}^{l-1}|z^{k+1}-z^k| \le b'(h/2)^{-\nu-1}n^{1/2}\,2d_G(x^1,x^2)$$

$$< 2n^{1/2}\,b'(h/2)^{-\nu-1}\delta \le \varepsilon, \quad \text{q.e.d.}$$

The proof of the implication (2.8) ⇒ (2.3) can be constructed in a similar way or simply using the symmetry argument.

Lemma 2.3. Let the kernel $K(x,y)$ be differentiable with respect to x on $(G\times G)\setminus\{x=y\}$ and satisfy conditions (2.1) and (2.7). Then, for any $u\in L^\infty(G)$ and any $x^1,x^2\in G$,

$$|(Tu)(x^1)-(Tu)(x^2)|$$

$$\le \text{const}\,\|u\|_{L^\infty(G)}\begin{Bmatrix} d_G(x^1,x^2) & , & \nu<n-1 \\ d_G(x^1,x^2)[1+|\log d_G(x^1,x^2)|], & & \nu=n-1 \\ [d_G(x^1,x^2)]^{n-\nu} & , & \nu>n-1 \end{Bmatrix} \tag{2.10}$$

where the constant does not depend on u and x^1,x^2.

Proof. The assertion of the Lemma is trivial if x^1 and x^2 belong to different connectivity components of G. Consider the case where x^1 and x^2 are in a common connectivity component. Denote $h=4d_G(x^1,x^2)$ and construct a polygonal path $[z^1,z^2,\dots,z^l]\subset G$ with nodes $x^1=z^1,z^2,\dots,z^l=x^2$ such that

$$\sum_{k=1}^{l-1}|z^{k+1}-z^k| \le 2d_G(x^1,x^2) = h/2.$$

We have

$$|(Tu)(x^1)-(Tu)(x^2)| = \Big|\int_G[K(x^1,y)-K(x^2,y)]u(y)dy\Big|$$

$$\le \|u\|_{L^\infty(G)}\Big(\int_{G\setminus(B(x^1,h)\cap B(x^2,h))}|K(x^1,y)-K(x^2,y)|dy+\sum_{j=1}^{2}\int_{G\cap B(x^j,h)}|K(x^j,y)|dy\Big)$$

Here, due to (2.1),

$$\int_{G\cap B(x^j,h)}|K(x^j,y)|dy \le \int_{B(x^j,h)}|x^j-y|^{-\nu}dy = b\sigma_n h^{n-\nu}/(n-\nu)$$

$$\le \text{const}[d_G(x^1,x^2)]^{n-\nu}, \quad j=1,2,$$

and this does not exceed the corresponding quantity in (2.10). Further, estimate the distance from $y \in G\setminus(B(x^1,h)\cap B(x^2,h))$ to our polygonal path:

$$|y-(tz^{k+1}-(1-t)z^k)| \ge \frac{1}{2}|y-x^1| \quad \text{for } 0\le t\le 1,\ k=1,\dots,l-1.$$

Using equality (2.9) and inequality (2.7) we estimate

$$\int_{G\setminus(B(x^1,h)\cap B(x^2,h))} |K(x^1,y)-K(x^2,y)|dy$$

$$\le b' \sum_{k=1}^{l-1}\sum_{i=1}^{n} |z_i^{k+1}-z_i^k| \int_{G\setminus(B(x^1,h)\cap B(x^2,h)} |y-(tz^{k+1}-(1-t)z^k)|^{-\nu-1}dy$$

$$\le b'\sqrt{n}\sum_{k=1}^{l-1}|z^{k+1}-z^k|\,2^{\nu+1}\int_{G\setminus B(x^1,h/2)}|y-x^1|^{-\nu-1}dy$$

$$\le ch\int_{h/2<|y|\le d}|y|^{-\nu-1}dy = ch\sigma_n\int_{h/2}^{d} r^{-\nu-2+n}dr \le c'\begin{Bmatrix} h & , \nu<n-1 \\ h(1+|\log h|), & \nu=n-1 \\ h^{n-\nu} & , \nu>n-1\end{Bmatrix}.$$

Recalling the meaning of h, estimate (2.10) follows. Lemma 2.3 is proved.

2.5. Weighted space $C^{m,\nu}(G)$. For a $\lambda\in\mathbb{R}$, introduce a weight function

$$w_\lambda(x) = \begin{Bmatrix} 1 & , \lambda<0 \\ [1+|\log\rho(x)|]^{-1} & , \lambda=0 \\ \rho(x)^\lambda & , \lambda>0 \end{Bmatrix}, \qquad x\in G \tag{2.11}$$

where $G\subset\mathbb{R}^n$ is an open bounded set and $\rho(x)=\inf_{y\in\partial G}|x-y|$ is the distance from x to ∂G. Let $m\in\mathbb{N}$, $\nu\in\mathbb{R}$, $\nu<n$. Define the space $C^{m,\nu}(G)$ as the collection of all m times continuously differentiable functions $u:G\to\mathbb{R}$ (or $\mathbb{C}$) such that

$$\|u\|_{m,\nu} \equiv \sum_{|\alpha|\le m}\sup_{x\in G}\left(w_{|\alpha|-(n-\nu)}(x)|D^\alpha u(x)|\right) < \infty.$$

In other words, a m times continuously differentiable function u on G belongs to $C^{m,\nu}(G)$ if the growth of its derivatives near the boundary can be estimated as follows:

$$|D^\alpha u(x)| \le \text{const}\begin{Bmatrix} 1 & , |\alpha|<n-\nu \\ 1+|\log\rho(x)| & , |\alpha|=n-\nu \\ \rho(x)^{n-\nu-|\alpha|} & , |\alpha|>n-\nu\end{Bmatrix}, \quad x\in G,\ |\alpha|\le m. \tag{2.12}$$

The space $C^{m,\nu}(G)$, equipted with the norm $\| \ \|_{m,\nu}$, is complete (a Banach space).

If $\nu<n-1$ then the first derivatives of a function $u\in C^{m,\nu}(G)$ are bounded on G, and as a consequence we have

$$|u(x^1)-u(x^2)| \le \text{const } d_G(x^1,x^2), \quad \forall x^1,x^2\in G. \tag{2.13}$$

Indeed, let x^1 and x^2 belong to a common connectivity component of G and let $[z^1,z^2,\dots,z^l]\subset G$ be polygonal path with nodes at $x^1=z^1,z^2,\dots,z^l=x^2$ such that

$$\sum_{k=1}^{l-1} |z^{k+1}-z^k| \le 2d_G(x^1,x^2).$$

We have

$$u(x^2)-u(x^1) = \sum_{k=1}^{l-1} u(z^{k+1})-u(z^k) = \sum_{k=1}^{l-1}\int_0^1 \frac{\partial}{\partial t} u(tz^{k+1}+(1-t)z^k)dt$$

$$= \sum_{k=1}^{l-1}\int_0^1 \sum_{i=1}^{n} \frac{\partial u(tz^{k+1}+(1-t)z^k)}{\partial x_i} dt(z_i^{k+1}-z_i^k).$$

Using the boundedness of the first derivatives of u we obtain

$$|u(x^2)-u(x^1)| \le c\sum_{k=1}^{l-1}\sum_{i=1}^{l-1} |z_i^{k+1}-z_i^k| \le c\sqrt{n}\sum_{k=1}^{l-1} |z^{k+1}-z^k| \le c'd_G(x^1,x^2),$$

q.e.d.

In case $\nu\ge n-1$ one can hope that, instead of (2.13), an inequality (cf.(2.10)

$$|u(x^2)-u(x^1)| \le \text{const} \begin{Bmatrix} d_G(x^1,x^2)(1+|\log d_G(x^1,x^2)|), & \nu=n-1 \\ (d_G(x^1,x^2))^{n-\nu} & , \ \nu>n-1 \end{Bmatrix} \tag{2.13'}$$

holds for $u\in C^{m,\nu}(G)$. This inequality is very useful when the convergence speed of approximate methods for weakly singular integral equations is analyzed. Unfortunately, the validity of (2.13') depends on the structure of G. We give a positive result and a counter-example.

Lemma 2.4. Assume that there are constants $\varepsilon_0>0$, $c>0$ and $c_0>0$ such that, for any $x^1,x^2\in G$ satisfying $d_G(x^1,x^2)\le\varepsilon_0$, there exist $\bar{x}^1,\bar{x}^2\in G$ satisfying the following conditions:

$$|\bar{x}^1-x^1| \le cd_G(x^1,x^2), \quad |\bar{x}^2-x^2| \le cd_G(x^1,x^2), \tag{2.14}$$

$$\rho(t\bar{x}^1+(1-t)x^1) \ge c_0|\bar{x}^1-x^1|t, \quad \rho(t\bar{x}^2+(1-t)x^2) \ge c_0|\bar{x}^2-x^2|t, \tag{2.15}$$

$$\rho(t\bar{x}^1+(1-t)\bar{x}^2) \ge c_0 d_G(x^1,x^2) \quad (0\le t\le 1). \tag{2.16}$$

Then, for $u \in C^{m,\nu}(G)$ with $\nu \geq n-1$, (2.13') holds.

Let us comment the conditions of the Lemma. If x^1 and x^2 selves satisfy (2.16) then we may put $\bar{x}^1 = x^1$, $\bar{x}^2 = x^2$, and (2.14), (2.15) will be trivially fulfilled. If x^1 and x^2 do not satisfy (2.16) we try to find points $\bar{x}^1$ and $\bar{x}^2$ in their neighborhoods defined by (2.14) such that $\bar{x}^1$ and $\bar{x}^2$ satisfy (2.16). The meaning of condition (2.15) is that there is a cone in G of a fixed vertex angle, with the vertex at x^1 or x^2, respectively, such that $\bar{x}^1$, respectively $\bar{x}^2$, lies on the axis of the cone. Figure 2.2 illustrates the placing of the points.

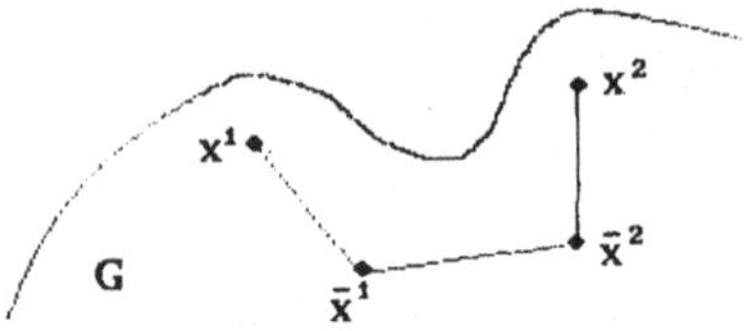

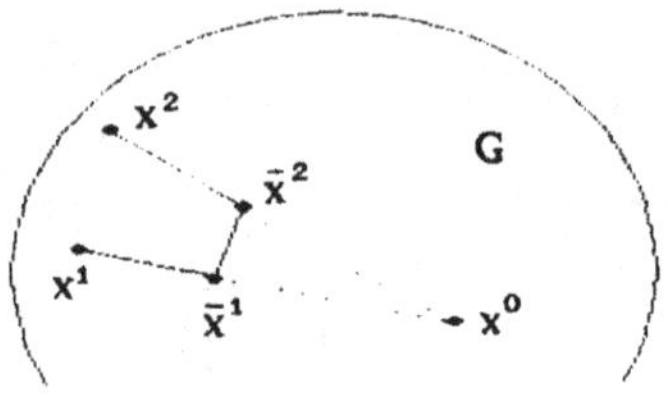

Fig. 2.2 Fig. 2.3

Note that conditions of Lemma 2.4 are fulfilled e.g. in case where G consists of a finite number of connectivity components whereby each of them is convex. In this case we can fix a point x^0 of a connectivity component and, for x^1 and x^2 from this connectivity component, choose $\bar{x}^1$ and $\bar{x}^2$ on the segments $[x^1, x^0]$ and $[x^2, x^0]$, respectively (see Figure 2.3).

Proof of Lemma 2.4. We have

$$|u(x^1)-u(x^2)| \leq |u(x^1)-u(\bar{x}^1)| + |u(\bar{x}^1)-u(\bar{x}^2)| + |u(\bar{x}^2)-u(x^2)|,$$

$$|u(x^k)-u(\bar{x}^k)| \leq \int_0^1 \sum_{i=1}^n \left| \frac{\partial u(tx^k+(1-t)\bar{x}^k)}{\partial x_i} \right| dt \ |x_i^k - \bar{x}_i^k|, \quad k=1,2,$$

$$|u(\bar{x}^1)-u(\bar{x}^2)| \leq \int_0^1 \sum_{i=1}^n \left| \frac{\partial u(t\bar{x}^1+(1-t)\bar{x}^2)}{\partial x_i} \right| dt \ |\bar{x}_i^1 - \bar{x}_i^2|$$

(cf. the proof of (2.13)). For $u \in C^{m,\nu}(G)$ with $\nu \geq n-1$, we have

$$\left| \frac{\partial u(x)}{\partial x_i} \right| \leq \text{const} \begin{Bmatrix} 1+|\log \rho(x)|, & \nu = n-1 \\ (\rho(x))^{n-\nu-1}, & \nu > n-1 \end{Bmatrix},$$

and together with (2.15), (2.16) we obtain

$$|u(x^k)-u(\bar{x}^k)| \leq c|x^k - \bar{x}^k| \int_0^1 \begin{Bmatrix} 1+|\log(c_0|x^k-\bar{x}^k|t)|, & \nu = n-1 \\ (c_0|x^k-\bar{x}^k|t)^{n-\nu-1}, & \nu > n-1 \end{Bmatrix} dt$$

$$\le c'\,|x^k-\bar{x}^k|\left\{\begin{array}{ll}1+|\log|x^k-\bar{x}^k||, & \nu=n-1\\ |x^k-\bar{x}^k|^{n-\nu-1}, & \nu>n-1\end{array}\right\},\quad k=1,2,$$

$$|u(\bar{x}^1)-u(\bar{x}^2)|\le c\,|\bar{x}^1-\bar{x}^2|\left\{\begin{array}{ll}1+|\log d_G(x^1,x^2)|, & \nu=n-1\\ (d_G(x^1,x^2))^{n-\nu-1}, & \nu>n-1\end{array}\right\}.$$

Now use (2.14) and a consequence of (2.14): $|\bar{x}^1-\bar{x}^2|\le(2c+1)d_G(x^1,x^2)$. The result is (2.13'). The proof of Lemma 2.4 is finished.

Let us present a <u>counter-example</u> showing that (2.13') fails in some cases. Let $G\subset\mathbb{R}^2$ be delimited by lines $x_2=\pm x_1^2$ and $x_1=1$ (see Figure 2.4). Then $\rho(x)\le x_1^2$ for $x=(x_1,x_2)\in G$ with $0<x_1<1/2$ whereby $\rho(x)x_1^{-2}\to 1$ for $x=(x_1,0)$, $x_1\to 0$. Consider the function $u(x)=x_1^{1/2}$, $x\in G$. Its derivatives

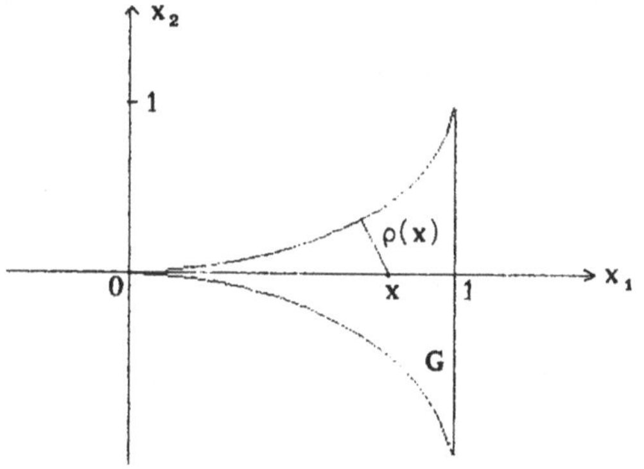

Fig. 2.4

$\partial u/\partial x_1=\frac{1}{2}x_1^{-1/2}$, $\partial^2u/\partial x_1^2=-\frac{1}{4}x_1^{-3/2}$ have singularities as $x_1\to 0$, $x\in G$. To state that $u\in C^{2,\nu}(G)$ we must find a ν $(n-1=1<\nu<2=n)$ such that (see (2.12))

$$|\partial^k u/\partial x_1^k|\le c\rho(x)^{2-\nu-k},\quad k=1,2.$$

Replacing $\rho(x)$ by x_1^2 we estabilish more strong inequalities

$$|\partial^k u/\partial x_1^k|\le c(x_1^2)^{2-\nu-k},\quad k=1,2,$$

with $\nu=5/4$. Thus, $u\in C^{2,5/4}(G)$. Now note that (2.13') fails for $u=x_1^{1/2}$. Indeed, for $x^1=(2t,0)$, $x^2=(t,0)$, (2.13') reads $(\sqrt{2}-1)t^{1/2}\le ct^{3/4}$ that is false for small $t>0$.

In this example, an important circumstance is that G has a zero angle and the cone condition is violated therefore.

2.6. Weighted space $C_a^{m,\nu}(G)$. Assume now that the boundary ∂G is piecewise smooth. Let $a(x)=(a_1(x),\ \ldots\ ,a_n(x))$, $x\in\bar{G}$, be a vector field of class $[C^m(\overline{G})]^n$. We construct this field to be tangent to ∂G (if possible) or to some part of ∂G (see Figure 2.5). Denote by Γ_a the part of ∂G to which

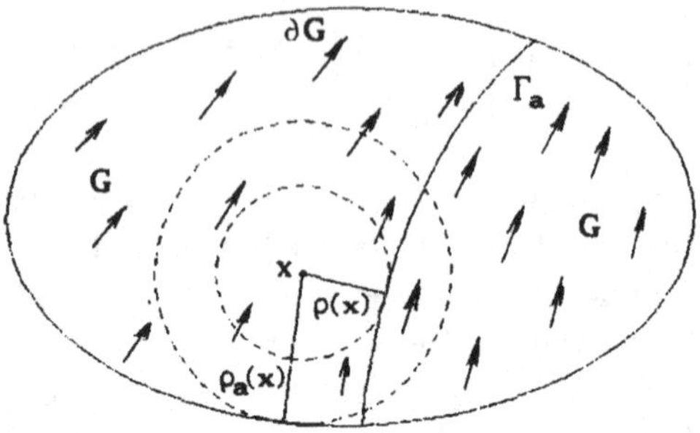

Fig. 2.5

a is tangent. More precisely, $x^0 \in \Gamma_a$ means that $x^0 \in \partial G$, $a(x^0) \neq 0$ and the vector $a(x^0)$ is orthogonal to the normal of each hypersurface at x^0 that is a part of ∂G and contains x^0.

We introduce the "tangential differentation" operators

$$\mathcal{L}_a^1 = \mathcal{L}_a = \sum_{j=1}^n a_j(x)\frac{\partial}{\partial x_j}, \quad \mathcal{L}_a^k = \mathcal{L}_a \mathcal{L}_a^{k-1}, \quad k = 2, \dots, m,$$

and the distance function (see Figure 2.5)

$$\rho_a(x) = \text{dist}(x, \partial G \setminus \Gamma_a) = \inf_{y \in \partial G, y \notin \Gamma_a} |x-y| \quad (x \in \overline{G})$$

setting $\rho_a(x) \equiv \text{diam}\, G$ if $\Gamma_a = \partial G$. Further, we introduce the weight functions (cf. (2.11))

$$w_{\lambda,a}(x) = \begin{cases} 1 & , \ \lambda < 0 \\ [1+\log\rho_a(x)]^{-1}, & \lambda = 0 \\ \rho_a(x)^\lambda & , \ \lambda > 0 \end{cases}, \quad x \in \overline{G}, \quad \lambda \in \mathbb{R},$$

and the semi-norm

$$|u|_{m,\nu,a} = \sum_{k=1}^m \sup_{x \in G} w_{k-(n-\nu),a}(x)\, |\mathcal{L}_a^k u(x)|.$$

We define the weighted space $C_a^{m,\nu}(G)$ to consist of all functions $u \in C^{m,\nu}(G)$ that have finite norm

$$\|u\|_{m,\nu,a} = \|u\|_{m,\nu} + |u|_{m,\nu,a}.$$

The space $C_a^{m,\nu}(G)$, equipted with the norm $\|\cdot\|_{m,\nu,a}$, is compete (a Banach space).

2.7. Exercises. 2.7.1. Let $G \subset \mathbb{R}^n$ be an open bounded set. For a subset $G' \subseteq G$, denote by

$$d_G\text{-}\operatorname{diam} G' = \sup_{x^1, x^2 \in G'} d_G(x^1, x^2)$$

its diameter with respect to the metric d_G. Prove the following assertion: G^* is compact with respect to d_G if and only if, for any $\varepsilon > 0$, there exists a partition of G into a finite number of disjoint sets $G_j \subseteq G$ of d_G-diameter not exceeding ε.

2.7.2. Introduce an analogue of condition (2.2) where the d_G-distance is replaced by the Euclidean distance:

$$\left.\begin{aligned} &\forall \varepsilon > 0 \quad \forall h > 0 \quad \exists \delta = \delta(\varepsilon, h) > 0: \quad x^1, x^2, y \in G, \quad |x^i - y| \geq h \quad (i = 1,2), \\ &|x^1 - x^2| < \delta \quad \Rightarrow \quad |K(x^1, y) - K(x^2, y)| < \varepsilon. \end{aligned}\right\} \qquad (2.2')$$

Prove that for $p > n/(n-\nu)$, the operator $T \in \mathcal{L}(L^p(G), C(\overline{G}))$ is compact provided that (2.1) and (2.2') hold.

2.7.3. As we showed in Section 2.4, condition (2.7) implies (2.2). In general, (2.7) does not imply (2.2') since (2.2') does not allow discontinuities of $K(x,y)$ on an inner boundary. Become convinced that (2.7) does not imply (2.2') even in the case illustrated on Figure 2.6 ($n = 2$; no inner boundary but a cuspidal point on ∂G).

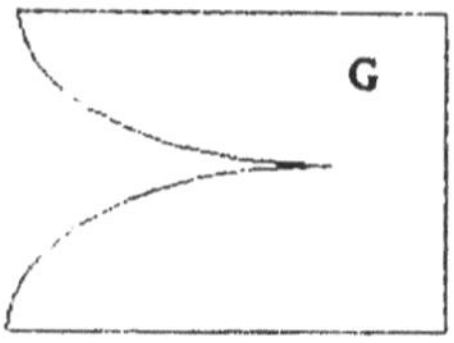

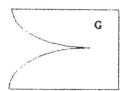

Fig. 2.6.

2.7.4. Prove the completeness of the space $C^{m,\nu}(G)$ and $C_a^{m,\nu}(G)$.

2.7.5. In general, $C^{m,\nu}(G)$ need not be imbedded into $BC(G^*)$. Let $G \subset \mathbb{R}^2$ be the region illustrated on Figure 2.4. Show that the function $u(x) = \cos(x_1^{-1/2})$ belongs to $C^{1,7/4}(G)$ but not to $BC(G^*)$.

2.7.6. Formulate sufficient conditions for the imbedding $C^{m,\nu}(G) \subset BC(G^*)$.

3. SMOOTHNESS OF THE SOLUTION

As a rule, a diagonal singularity of the kernel of an integral equation $u = Tu + f$ causes singularities of the derivatives of the solution at the boundary of the domain. The purpose of this Chapter is to characterize those singularities.

3.1. The class of weakly singular kernels. Consider the integral equation of the second kind

$$u(x) = \int_G K(x,y)u(y)dy + f(x), \quad x \in G, \tag{3.1}$$

where $G \subset \mathbb{R}^n$ is an open bounded set and $K(x,y)$ is a real or complex valued kernel; later we shall discuss the case of operator valued kernels, too. A differentiation of a weakly singular kernel increases the order of the singularity, e.g. $D_x^\alpha\ |x-y|^{-\nu}$ $(\nu > 0)$ behaves as $|x-y|^{-\nu-|\alpha|}$; on the other hand, $(\partial/\partial x_1 + \partial/\partial y_1)|x-y|^{-\nu} = 0$. These observations motivate to introduce the smoothness assumption about the kernel in the following form:

the kernel $K(x,y)$ is m times $(m \ge 1)$ continuously differentiable on $(G\times G)\setminus\{x=y\}$ and there exists a real number $\nu \in (-\infty, n)$ such that the estimate

$$\left|\left(\frac{\partial}{\partial x_1}\right)^{\alpha_1} \cdots \left(\frac{\partial}{\partial x_n}\right)^{\alpha_n}\left(\frac{\partial}{\partial x_1}+\frac{\partial}{\partial y_1}\right)^{\beta_1} \cdots \left(\frac{\partial}{\partial x_n}+\frac{\partial}{\partial y_n}\right)^{\beta_n} K(x,y)\right|$$

$$\le b \begin{Bmatrix} 1 & , & \nu+|\alpha|<0 \\ 1+|\log|x-y|| & , & \nu+|\alpha|=0 \\ |x-y|^{-\nu-|\alpha|} & , & \nu+|\alpha|>0 \end{Bmatrix}, \quad b = \text{const} \quad (x,y\in G) \tag{3.2}$$

holds for all multi-indeces $\alpha = (\alpha_1, \dots, \alpha_n) \in \mathbb{Z}_+^n$ and $\beta = (\beta_1, \dots, \beta_n) \in \mathbb{Z}_+^n$ with $|\alpha| + |\beta| \le m$.

Here the following usual conventions are adopted:

$$|\alpha| = \alpha_1 + \dots + \alpha_n \quad \text{for} \quad \alpha = (\alpha_1, \dots, \alpha_n) \in \mathbb{Z}_+^n,$$

$$|x| = (x_1^2 + \dots + x_n^2)^{1/2} \quad \text{for} \quad x = (x_1, \dots, x_n) \in \mathbb{R}^n.$$

Putting $|\alpha| = |\beta| = 0$, (3.2) yields

$$|K(x,y)| \le b \begin{Bmatrix} 1 & , & \nu<0 \\ 1+|\log|x-y|| & , & \nu=0 \\ |x-y|^{-\nu} & , & \nu>0 \end{Bmatrix}.$$

Thus, K(x,y) may have a weak singularity ($\nu < n$). In case $\nu<0$, the kernel is bounded but its derivatives may be singular.

Let us denote by $D_x^\alpha D_{x+y}^\beta$ the differential operator which is applied to K(x,y) in the left hand side of (3.2). The assymmetry of (3.2) with respect to x and y is only seeming; actually, using the equality $\frac{\partial}{\partial y_i}=\left(\frac{\partial}{\partial x_i}+\frac{\partial}{\partial y_i}\right)-\frac{\partial}{\partial x_i}$ we can deduce from (3.2) a similar estimate for $D_y^\alpha D_{x+y}^\beta K(x,y)$:

$$|D_y^\alpha D_{x+y}^\beta K(x,y)| \le b' \begin{Bmatrix} 1 & , & \nu+|\alpha|<0 \\ 1+|\log|x-y|| & , & \nu+|\alpha|=0 \\ |x-y|^{-\nu-|\alpha|} & , & \nu+|\alpha|>0 \end{Bmatrix}, \quad |\alpha|+|\beta|\le m. \tag{3.2'}$$

From (3.2) and (3.2') with $|\alpha|=1$, $|\beta|=0$ we obtain that K(x,y) satisfies conditions (2.2) and (2.3). Consequently (see Lemma 2.2), under condition (3.2), the integral operator T with kernel K(x,y) is compact from $L^\infty(G)$ to $BC(G^*)$.

For subsequent references, we present some inequalities that follow from (3.2). Denoting $K_\beta(x,y)=D_{x+y}^\beta K(x,y)$, we have for $|\alpha|+|\beta|\le m$, $x\in G$, $r>0$ the estimates

$$\int_{G\cap B(x,r)} |D_x^\alpha K_\beta(x,y)|\,dy \le c_{n,\nu,b} \begin{Bmatrix} r^n & , & \nu+|\alpha|<0 \\ r^n(1+|\log r|), & & \nu+|\alpha|=0 \\ r^{n-\nu-|\alpha|} & , & 0<\nu+|\alpha|<n \end{Bmatrix}, \tag{3.3}$$

$$\int_{G\setminus B(x,r)} |D_x^\alpha K_\beta(x,y)|\,dy \le c_{n,\nu,b,d} \begin{Bmatrix} 1 & , & \nu+|\alpha|<n \\ 1+|\log r|, & & \nu+|\alpha|=n \\ r^{n-\nu-|\alpha|}, & & \nu+|\alpha|>n \end{Bmatrix} \tag{3.4}$$

where the constants $c_{n,\nu,b}$ and $c_{n,\nu,b,d}$ depend only on n,ν,b and n,ν,b,d respectively where d is the diameter of G. The proof can be obtained by a direct calculations. We demonstrate the techniques proving (3.4) in case $\nu+|\alpha|>0$:

$$\int_{G\setminus B(x,r)} |D_x^\alpha K_\beta(x,y)|\,dy \le b \int_{r<|x-y|<d} |x-y|^{-\nu-|\alpha|}dy = b \int_{r<|y|<d} |y|^{-\nu-|\alpha|}dy =$$

$$= b\int_r^d \rho^{-\nu-|\alpha|}\sigma_n\rho^{n-1}d\rho = b\sigma_n \begin{Bmatrix} \frac{1}{n-\nu-|\alpha|}\left(d^{n-\nu-|\alpha|}-r^{n-\nu-|\alpha|}\right), & \nu+|\alpha|<n \\ \log d-\log r & , \ \nu+|\alpha|=n \\ \frac{1}{\nu+|\alpha|-n}\left(r^{n-\nu-|\alpha|}-d^{n-\nu-|\alpha|}\right), & \nu+|\alpha|>n \end{Bmatrix}$$

where σ_n is the area of the unit sphere of $\mathbb{R}^n$; we assumed that $r \le d$ (otherwise $G \setminus B(x,r)$ is void). Now (3.4) follows. The proof of (3.3) is partly done in Section 2.2.

Let us discuss some examples of the kernels satisfying (3.2). First, the kernel $K(x,y) = a(x,y)|x-y|^{-\nu}$ $(0<\nu<n)$ satisfies (3.2) if $a(x,y)$ is m times continuously differentiable on $(G\times G)\setminus\{x=y\}$ and

$$|D_x^\alpha D_{x+y}^\beta a(x,y)| \le b'|x-y|^{-|\alpha|} \quad (|\alpha|+|\beta| \le m), \quad b' = \text{const}.$$

In Section 1.3, the kernel of integral equation (1.5) has the form $K(x,y) = -c_n a(x)|x-y|^{-n+2}$, $n \ge 3$. Condition (3.2) is fulfilled if $a \in BC^m(G)$, i.e. $a \in C^m(G)$ and its derivatives up to the m-th order are bounded on G.

The kernel $K(x,y)=a(x,y)\log|x-y|$ satisfies (3.2) with $\nu=0$ if $a(x,y)$ is m times continuously differentiable on $(G\times G)\setminus\{x=y\}$ and the derivatives are bounded (more general conditions on the growth of the derivatives of a can be formulated, too). In Section 1.4 we dealed with integral equations having the kernel $K(x,y)=(2\pi)^{-1}a(x)\log|x-y|$, $n=2$. Condition (3.2) is fulfilled for it if $a \in BC^m(G)$. Thus, $\nu=n-2$ in integral equations considered in Sections 1.3 and 1.4.

An example with $\nu=n-1$ is given by Peierls kernel (see Section 1.7)

$$K(x,y) = \frac{1}{4\pi}\,\sigma_0(x)\exp(-\tau(x,y))|x-y|^{-2}, \quad n=3,$$

where $\tau(x,y)$ is given by formula (1.14); $G \subset \mathbb{R}^3$ is assumed to be convex in this example. Condition (3.2) is fulfilled if $\sigma, \sigma_0 \in BC^m(G)$. The cases $\nu=n-2$ and $\nu=n-1$ seem to occur most frequently in the practice.

Now we extend our view to equation (3.1) allowing E valued functions $u: G \to E$, $f: G \to E$ and $\mathcal{L}(E,E)$ valued kernels $K: G\times G \to \mathcal{L}(E,E)$ where E is a Banach space; in (3.2) $|D_x^\alpha D_{x+y}^\beta K(x,y)|$ then designates the norm of the linear operator whereby the continuity of the derivatives of K may be understood as the continuity in the uniform operator topology or in the strong operator topology. This enables to treat a system of integral equations

$$u_i(x) = \sum_{j=1}^{l} \int_G K_{ij}(x,y)\,u_j(y)\,dy + f_i(x), \qquad i=1,\dots,l,$$

as equation (3.1) with $\mathcal{L}(\mathbb{R}^l,\mathbb{R}^l)$ valued kernel $K(x,y) = (K_{ij}(x,y))_{i,j=1}^{l}$. For the matrix function $K(x,y)$, condition (3.2) means that every partial kernel $K_{ij}(x,y)$, $i,j=1,\dots,l$, satisfies (3.2).

Perhaps it is more unexpected that the general radiation transfer equation (1.15) can be also fitted into the framework of (3.1)-(3.2). Assume that $G \subset \mathbb{R}^3$ is convex and the functions $\sigma_0, \sigma: \overline{G} \to \mathbb{R}$, $g: \overline{G}\times S\times S \to \mathbb{R}$ are sufficiently smooth. We can treat (1.15) as equation (3.1) with $\mathcal{L}(C^m(S), C^m(S))$

valued kernel $K(x,y)$ defined by the formula

$$(K(x,y)v)(s) = \frac{\sigma_0(x)}{4\pi} g\left(x,s,\frac{x-y}{|x-y|}\right) \exp(-\tau(x,y))|x-y|^{-2} v\left(\frac{x-y}{|x-y|}\right),$$

$$s\in S, \quad v\in C^m(S).$$

The differentiability of $K(x,y)$ and the check of (3.2) need in a special analysis. We are not ready to present it here.

3.2. Main results and comments. Now we are ready to formulate the main results of the chapter; after some prelimenaries the proofs are given in Sections 3.5 and 3.8.

We refer to Sections 2.5 and 2.6 for the definitions of the weighted spaces $C^{m,\nu}(G)$ and $C_a^{m,\nu}(G)$.

Theorem 3.1. Let $G\subset\mathbb{R}^n$ be an open bounded set, $f\in C^{m,\nu}(G)$ and let the kernel $K(x,y)$ satisfy condition (3.2). If integral equation (3.1) has a solution $u\in L^\infty(G)$ (or even $u\in L(G)$) then $u\in C^{m,\nu}(G)$.

Theorem 3.2. Let $G\subset\mathbb{R}^n$ be an open bounded set with piecewise smooth boundary ∂G, and let $a(x)=(a_1(x), \ldots , a_n(x))$, $x\in\overline{G}$, be a vector field of class $[C^m(\overline{G})]^n$. Suppose that $f\in C_a^{m,\nu}(G)$ and the kernel $K(x,y)$ satisfies condition (3.2). If integral equation (3.1) has a solution $u\in L^\infty(G)$ (or $u\in L(G)$) then $u\in C_a^{m,\nu}(G)$.

The main qualitative consequence of Theorems 3.1 and 3.2 is that if f and ∂G are smooth then the tangential derivatives of a solution behave essentially better than the normal (or other non-tangential) derivatives.

Remark 3.1. Theorem 3.1 holds without any assumption about ∂G.

Remark 3.2. In Theorems 3.1 and 3.2, we have not assumed the uniqueness of the solution to (3.1). In case $f=0$, we immediately get that the eigenfunctions of the integral operator T with the kernel $K(x,y)$ on G satisfying (3.2) belong to $C^{m,\nu}(G)$, respectively, $C_a^{m,\nu}(G)$. By induction, we get that the generalized eigenfunctions also belong to $C^{m,\nu}(G)$, respectively, $C_a^{m,\nu}(G)$, e.g.

$$\lambda_0 u_0 = Tu_0\ ,\ \lambda_0\neq 0,\ u_0\neq 0 \quad\Rightarrow\quad u_0\in C^{m,\nu}(G),$$

$$\lambda_0 u_i = Tu_i + u_{i-1}\ ,\ u_{i-1}\in C^{m,\nu}(G) \quad\Rightarrow\quad u_i\in C^{m,\nu}(G),\ i=1,2,\ldots,l.$$

Thus, the whole generalized eigenspace of T corresponding to an eigenvalue λ_0 belongs to $C^{m,\nu}(G)$, respectively, $C_a^{m,\nu}(G)$.

Remark 3.3. As we shall see later, in general, a solution to integral equation (3.1) does not improve its properties near ∂G, remaining only in $C^{m,\nu}(G)$, even if ∂G is of class C^∞ and $f\in C^\infty(\overline{G})$. More precisely, for any n and ν $(\nu<n)$ there are kernels $K(x,y)$ satisfying (3.2) and such that (3.1)

is uniquely solvable and, for a suitable $f \in C^{\infty}(\bar{G})$, the normal derivatives of order k of the solution to (3.1) behave near ∂G as $\log \rho(x)$ if $k = n-\nu$ and as $\rho(x)^{n-\nu-k}$ for $k > n-\nu$. On the other hand, there are special kernels $K(x,y)$ such that all derivatives behave better if ∂G is smooth and $f \in C^m(G^*)$ (see Section 3.11).

Remark 3.4. Theorems 3.1 and 3.2 remain true for E-valued functions $f,u: G \to E$ and $\mathcal{L}(E,E)$-valued kernels $K: G\times G \to \mathcal{L}(E,E)$ where E is a Banach space.

See the discussion in the end of Section 3.1. The extension of the definitions of the weighted spaces $C^{m,\nu}(G)$ and $C_a^{m,\nu}(G)$ to the case of E-valued functions is quite immediate. For instance, $C^{m,\nu}(G)$ consists of all m times continuously differentiable functions $u: G \to E$ such that

$$\|u\|_{m,\nu} \equiv \sum_{|\alpha| \le m} \sup_{x \in G} (w_{|\alpha|-(n-\nu)}(x)\, |D^{\alpha}u(x)|) < \infty$$

where $|\cdot|$ means the norm of E.

We present during Sections 3.3–3.8 the proofs of Theorems 3.1 and 3.2 which hold in the case of the extended understanding of equation (3.1) and condition (3.2). But the reader may restrict himself to case $E = \mathbb{R}$ (or $\mathbb{C}$) which is supported by the designations.

3.3. Differentiation of weakly singular integral. We use the following standard desingations for an open ball, a closed ball and a sphere in $\mathbb{R}^n$:

$$B(x,r) = \{y \in \mathbb{R}^n : |x-y| < r\},$$

$$\bar{B}(x,r) = \{y \in \mathbb{R}^n : |x-y| \le r\},$$

$$S(x,r) = \{y \in \mathbb{R}^n : |x-y| = r\}.$$

Lemma 3.1. Let $u \in L^{\infty}(G) \cap C^1(G)$ and let $K(x,y)$ satisfy (3.2) with $m = 1$. Then the function $\int_G K(x,y)u(y)dy$ is continuously differentiable on G, and for any $x \in G$ and $\bar{B}(x,r) \subset G$, the following differentiation formula holds:

$$\frac{\partial}{\partial x_i} \int_G K(x,y)u(y)dy = \int_{G\setminus B(x,r)} \frac{\partial K(x,y)}{\partial x_i} u(y)dy$$

$$+ \int_{B(x,r)} \Big(\frac{\partial}{\partial x_i} + \frac{\partial}{\partial y_i}\Big)\big[K(x,y)u(y)\big]dy + \int_{S(x,r)} K(x,y)u(y)\, \frac{x_i - y_i}{r}\, dS_y. \tag{3.5}$$

Proof. Let us denote $\varkappa(x,y) = K(x,y)u(y)$. We shall prove that

$$\frac{\partial}{\partial x_i} \int_{B(x,r)} \varkappa(x,y)dy = \int_{B(x,r)} \Big(\frac{\partial}{\partial x_i} + \frac{\partial}{\partial y_i}\Big)\varkappa(x,y)dy, \tag{3.6}$$

$$\frac{\partial}{\partial x_i}\int_{G\setminus B(x,r)} \varkappa(x,y)\,dy = \int_{G\setminus B(x,r)} \frac{\partial \varkappa(x,y)}{\partial x_i}\,dy + \int_{S(x,r)} \varkappa(x,y)\,\frac{x_i - y_i}{r}\,dS_y. \quad (3.7)$$

Formula (3.5) is an immediate consequence of (3.6) and (3.7). It can be seen from (3.5) that $(\partial/\partial x_i)\int_G K(x,y)u(y)\,dy$ is continuous at x.

Denote

$$e^i = (\underbrace{0,\dots,0}_{i-1},1,\underbrace{0,\dots,0}_{n-i})$$

Consider the difference quotient which corresponds to the derivative $(\partial/\partial x_i)\int_{B(x,r)}\varkappa(x,y)\,dy$:

$$\frac{1}{h}\Big\{\int_{B(x+he^i,r)} \varkappa(x+he^i,y)\,dy - \int_{B(x,r)} \varkappa(x,y)\,dy\Big\}$$

$$= \int_{B(x,r)} \frac{\varkappa(x+he^i,y+he^i)-\varkappa(x,y)}{h}\,dy$$

$$\to \int_{B(x,r)} \Big(\frac{\partial}{\partial x_i}+\frac{\partial}{\partial y_i}\Big)\varkappa(x,y)\,dy \quad \text{as} \quad h\to 0.$$

This yields (3.6). The last convergence can be argued using the Lebesque Theorem about the limiting process under the integral sign: first, the integrand converges almost everywhere on $B(x,r)$,

$$\frac{1}{h}\Big[\varkappa(x+he^i,\ y+he^i)-\varkappa(x,y)\Big] \to \Big(\frac{\partial}{\partial x_i}+\frac{\partial}{\partial y_i}\Big)\varkappa(x,y) \quad \text{for} \quad y\neq x;$$

secondly, due to (3.2)

$$\Big|\frac{1}{h}\Big[\varkappa(x+he^i,y+he^i)-\varkappa(x,y)\Big]\Big| = \Big|\frac{1}{h}\int_0^1 \frac{d}{dt}\,\varkappa(x+the^i,y+the^i)\,dt\Big|$$

$$= \Big|\int_0^1 \Big(\frac{\partial}{\partial x_i}+\frac{\partial}{\partial y_i}\Big)\varkappa(x+the^i,y+the^i)\,dt\Big| \le \int_0^1 \Big|\Big(\frac{\partial}{\partial x_i}+\frac{\partial}{\partial y_i}\Big)\varkappa(x+the^i,y+the^i)\Big|\,dt$$

$$\le \text{const}\,|x-y|^{-\nu},$$

i.e. the integrands have a common majorant $\text{const}\,|x-y|^{-\nu}$ which is an integrable function, $\int_{B(x,r)}|x-y|^{-\nu}dy<\infty$.

Consider the difference quotient which corresponds to the derivative $(\partial/\partial x_i)\int_{G\setminus B(x,r)}\varkappa(x,y)\,dy$:

$$\frac{1}{h}\Big\{\int_{G\setminus B(x+he^{i},r)} \varkappa(x+he^{i},y)\,dy - \int_{G\setminus B(x,r)} \varkappa(x,y)\,dy\Big\}$$

$$= \int_{G\setminus B(x,r)} \frac{\varkappa(x+he^{i},y)-\varkappa(x,y)}{h}\,dy + \frac{1}{h}\Big\{\int_{G\setminus B(x+he^{i},r)} - \int_{G\setminus B(x,r)}\Big\}\varkappa(x+he^{i},y)\,dy$$

$$= \int_{G\setminus B(x,r)} \frac{\varkappa(x+he^{i},y)-\varkappa(x,y)}{h}\,dy + \frac{1}{h}\Big\{\int_{B(x,r)} - \int_{B(x+he^{i},r)}\Big\}\varkappa(x+he^{i},y)\,dy.$$

As $h\to 0$, we obtain formula (3.7). Note that here $|x-y|\geq r$, thus, due to (3.2) the integrands are bounded and we have no problems with the convergence of the integrals. Figure 3.1 explains some details concerning the entry of the surface integral over the sphere $S(x,r)$ as $h\to 0$. The domain of integration

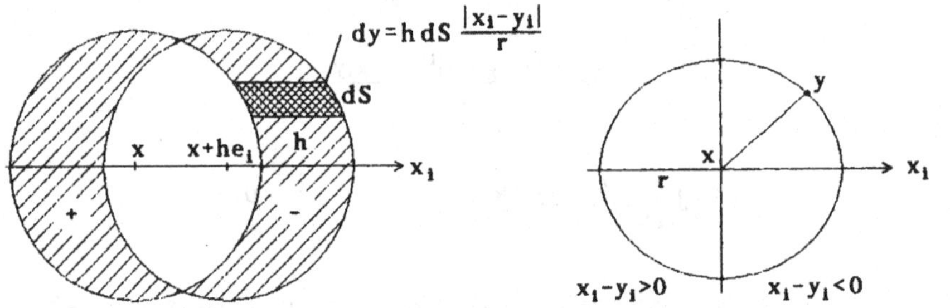

Fig. 3.1

in $\int_{B(x,r)} - \int_{B(x+he^{i},r)}$ consists of two "crescents", on one crescent the integral has sign "+", on another one the sign "-". The same signs has x_i-y_i for $y\in S(x,r)$. The proof of Lemma 3.1 is completed.

We add the formula

$$\frac{\partial}{\partial x_i}\int_{S(x,r)} \varkappa(x,y)\,dS_y = \int_{S(x,r)} \Big(\frac{\partial}{\partial x_i}+\frac{\partial}{\partial y_i}\Big)\varkappa(x,y)\,dS_y, \qquad \bar{B}(x,r)\subset G, \tag{3.8}$$

which can be proved in a similar way as (3.6). Now we are in a position to differentiate (3.5) once more using formulas (3.6)-(3.8). Assuming that $u\in L^{\infty}(G)\cap C^{2}(G)$ and (3.2) is fulfilled with $m=2$ we obtain

$$\frac{\partial^2}{\partial x_j\partial x_i}\int_G K(x,y)u(y)\,dy = \int_{G\setminus B(x,r)} \frac{\partial^2 K(x,y)}{\partial x_j\partial x_i}\,u(y)\,dy$$

$$+ \int_{B(x,r)} \Big(\frac{\partial}{\partial x_i}+\frac{\partial}{\partial y_i}\Big)\Big(\frac{\partial}{\partial x_j}+\frac{\partial}{\partial y_j}\Big)\big[K(x,y)u(y)\big]\,dy +$$

$$+\int_{S(x,r)} \frac{\partial K(x,y)}{\partial x_i} u(y) \frac{x_j - y_j}{r} dS_y + \int_{S(x,r)} \left[\left(\frac{\partial}{\partial x_j} + \frac{\partial}{\partial y_j}\right)(K(x,y)u(y))\right] \frac{x_i - y_i}{r} dS_y . \quad (3.9)$$

An equivalent formula will be obtained if we differentiate first with respect to x_j and then to x_i. We notice that the appereance of (equivalent) differentation formulas depends on the order of differentiations.

The following results can be proved by induction.

Lemma 3.2. Let $u \in L^\infty(G) \cap C^m(G)$ and let $K(x,y)$ satisfy (3.2). Then the function $\int_G K(x,y)u(y)dy$ is m times continuously differentiable on G, and, for any $x \in G$ with $\bar{B}(x,r) \subset G$, the following differentiation formula holds:

$$D_x^\alpha \int_G K(x,y)u(y)dy = \int_{G \setminus B(x,r)} D_x^\alpha K(x,y)u(y)dy + \int_{B(x,r)} D_{x+y}^\alpha [K(x,y)u(y)]dy$$

$$+\sum_{j=0}^{k-1} \int_{S(x,r)} \left[D_x^{\alpha - \alpha^{j+1}} D_{x+y}^{\alpha^j} K(x,y)u(y)\right] \frac{x_{i_{j+1}} - y_{i_{j+1}}}{r} dS_y , \qquad |\alpha| = k \le m, \ (3.10)$$

where $\alpha^0 = 0$ and the procession of multi-indices $\alpha^1, \alpha^2, \dots, \alpha^k = \alpha$ with

$$\alpha^1 = e^{i_1}, \quad \alpha^j = \alpha^{j-1} + e^{i_j} \quad (j = 2, \dots, k)$$

gives the order of differentations: $\partial/\partial x_{i_1}$ is computed first, then $(\partial/\partial x_{i_2})(\partial/\partial x_{i_1})$ and so on.

Now we present a reformulation of formula (3.5) to the case where G or its part $\Omega \subseteq G$ has a piecewise smooth boundary.

Lemma 3.3. Let $\Omega \subseteq G$ be a domain with a piecewise smooth boundary $\partial\Omega$. Let $u \in C(\bar{\Omega}) \cap C^1(\Omega)$, $\partial u/\partial x_i \in L(\Omega)$ and let $K(x,y)$ satisfy (3.2) with $m=1$. Then, for $x \in \Omega$,

$$\frac{\partial}{\partial x_i} \int_\Omega K(x,y)u(y)dy = \int_\Omega \left(\frac{\partial K(x,y)}{\partial x_i} + \frac{\partial K(x,y)}{\partial y_i} \right) u(y)dy + \int_\Omega K(x,y) \frac{\partial u(y)}{\partial y_i} dy$$

$$+ \int_{\partial\Omega} K(x,y)u(y)\,\omega_i(y)\,dS_y \quad (3.11)$$

where $\omega(y) = (\omega_1(y), \dots, \omega_n(y))$ is the unit inner normal to $\partial\Omega$ at $y \in \partial\Omega$.

Proof. We have (cf.(3.5)), for $x \in \Omega$ with $\bar{B}(x,r) \subset \Omega$

$$\frac{\partial}{\partial x_i} \int_\Omega K(x,y)u(y)dy = \int_{\Omega \setminus B(x,r)} \frac{\partial K(x,y)}{\partial x_i} u(y)dy$$

$$+ \int_{B(x,r)} \left(\frac{\partial}{\partial x_i} + \frac{\partial}{\partial y_i}\right)[K(x,y)u(y)]dy + \int_{S(x,r)} K(x,y)u(y) \frac{x_i - y_i}{r} dS_y$$

$$= \int_\Omega \left(\frac{\partial}{\partial x_i} + \frac{\partial}{\partial y_i}\right)[K(x,y)u(y)]dy - \int_{\Omega\setminus B(x,r)} \frac{\partial[K(x,y)u(y)]}{\partial y_i}dy$$

$$+ \int_{S(x,r)} K(x,y)u(y)\frac{x_i - y_i}{r}\,dS_y$$

Integrating by parts we find

$$-\int_{\Omega\setminus B(x,r)} \frac{\partial[K(x,y)u(y)]}{\partial y_i}dy = \int_{\partial\Omega} K(x,y)u(y)\,\omega_i(y)\,dS_y - \int_{S(x,r)} K(x,y)u(y)\frac{x_i - y_i}{r}\,dS_y;$$

note that $(x-y)/r$ is the unit outer normal to $S(x,r)$ as a part of the boundary to $\Omega\setminus B(x,r)$. The result is (3.11). Lemma 3.3 is proved.

Formulas for higher order derivatives can be presented, too.

3.4. Estimates of the derivatives of Tu. Let us denote

$$\|u\|_0 = \|u\|_{L^\infty(G)} = \sup_{x\in G}|u(x)|.$$

We refer to Section 2.5 for the definitions of the weight functions w_λ and the norm $\|.\|_{m,\nu}$ of the space $C^{m,\nu}(G)$.

<u>Lemma</u> **3.4**. Suppose that condition (3.2) for the kernel $K(x,y)$ holds. Then, for any $u \in C^{m,\nu}(G)$, the function

$$v(x) = (Tu)(x) = \int_G K(x,y)u(y)dy$$

also belongs to $C^{m,\nu}(G)$. Moreover,

$$\|v\|_0 \le c_0\|u\|_0, \quad c_0 = c_{n,\nu,b}\begin{Bmatrix} d^n & , \nu<0 \\ d^n(1+|\log d|), & \nu=0 \\ d^{n-\nu} & , \nu>0 \end{Bmatrix}, \quad d = \operatorname{diam} G, \tag{3.12}$$

and, for $x\in G$ and $1\le|\alpha|\le m$, the pointwise estimate

$$w_{|\alpha|-(n-\nu)}(x)\,|D^\alpha v(x)|$$

$$\le \begin{Bmatrix} c_1\|u\|_0 + c_2\rho(x)^{\min\{1,n-\nu\}}\|u\|_{|\alpha|,\nu} & \text{if } \nu \text{ is a fraction} \\ c_1\|u\|_0 + c_2\rho(x)(1+|\log\rho(x)|)\|u\|_{|\alpha|,\nu} & \text{if } \nu \text{ is an integer} \end{Bmatrix} \tag{3.13}$$

holds where the constants $c_i = c_i(b,n,\nu,d)$, $i=1,2$, are independent of $x\in G$ and $u\in C^{m,\nu}(G)$ and remain bounded as $d\to 0$.

<u>Proof</u>. Estimate (3.12) follows immediately from (3.3). Further, due to Lemma 3.2, the function v is m times continuously differentiable on G and its derivatives can be represented according to formula (3.10). Using the

Leibniz rule

$$D^{\alpha}(u_1 u_2) = \sum_{0\le\beta\le\alpha} \binom{\alpha}{\beta}\left(D^{\alpha-\beta}u_1\right)(D^{\beta}u_2)$$

and recalling the designation $K_{\alpha}(x,y) = D^{\alpha}_{x+y}K(x,y)$ we rewrite (3.10) in the form

$$D^{\alpha}v(x) = \int_{G\setminus B(x,r)} D^{\alpha}_{x} K(x,y)u(y)dy + \sum_{0\le\beta\le\alpha}\binom{\alpha}{\beta}\int_{B(x,r)} K_{\alpha-\beta}(x,y)D^{\beta}u(y)dy$$

$$+ \int_{S(x,r)} [D^{\alpha-\alpha^1}_{x} K(x,y)]u(y)\,\frac{x_{i_1}-y_{i_1}}{r}\,dS_y \tag{3.14}$$

$$+\sum_{j=1}^{|\alpha|-1}\ \sum_{0\le\beta\le\alpha^j}\binom{\alpha^j}{\beta}\int_{S(x,r)}\left[D^{\alpha-\alpha^{j+1}}_{x}K_{\alpha^j-\beta}(x,y)\right]D^{\beta}u(y)\,\frac{x_{i_{j+1}}-y_{i_{j+1}}}{r}\,dS_y$$

(from the sum of the surfase integrals we raised the term with $j=0$ separately). Recall that $|\alpha^j|=j$. We may use (3.14) with any $r>0$ such that $\bar{B}(x,r)\subset G$, i.e. $0<r<\rho(x)$. We put $r=\rho(x)/2$. Then, for $y\in\bar{B}(x,r)$,

$$\frac{1}{2}\,\rho(x) \le \rho(y) \le \frac{3}{2}\,\rho(x),$$

and $w_{\lambda}(x)$ and $w_{\lambda}(y)$ are of the same order, e.g.

$$\left(\frac{1}{2}\right)^{\lambda} w_{\lambda}(x) \le w_{\lambda}(y) \le \left(\frac{3}{2}\right)^{\lambda} w_{\lambda}(x), \quad \lambda>0. \tag{3.15}$$

Let us estimate the terms on the right side of (3.14). First, due to (3.4),

$$\left|\int_{G\setminus B(x,r)} D^{\alpha}_{x} K(x,y)u(y)dy\right|$$

$$\le c\begin{Bmatrix} 1 & , & \nu+|\alpha|<n \\ 1+|\log r|, & & \nu+|\alpha|=n \\ r^{n-\nu-|\alpha|}, & & \nu+|\alpha|>n\end{Bmatrix}\|u\|_0 \le c'\begin{Bmatrix} 1 & , & |\alpha|<n-\nu \\ 1+|\log\rho(x)|, & & |\alpha|=n-\nu \\ \rho(x)^{n-\nu-|\alpha|}, & & |\alpha|>n-\nu\end{Bmatrix}\|u\|_0$$

$$= c'[w_{|\alpha|-(n-\nu)}(x)]^{-1}\|u\|_0.$$

The area of $S(x,r)$ is equal to $\sigma_n r^{n-1}$, and with the help of (3.2) we find

$$\left|\int_{S(x,r)} [D^{\alpha-\alpha^1}_{x} K(x,y)]u(y)\,\frac{x_{i_1}-y_{i_1}}{r}\,dS_y\right|$$

$$\le c r^{n-1}\begin{Bmatrix} 1 & , & \nu+|\alpha|-1<0 \\ 1+|\log r|, & & \nu+|\alpha|-1=0 \\ r^{-\nu-|\alpha|+1}, & & \nu+|\alpha|-1>0\end{Bmatrix}\|u\|_0 \le c'\begin{Bmatrix} \rho(x)^{n-1} & , & |\alpha|<1-\nu \\ \rho(x)^{n-1}(1+|\log\rho(x)|), & & |\alpha|=1-\nu \\ \rho(x)^{n-\nu-|\alpha|} & , & |\alpha|>1-\nu\end{Bmatrix}\|u\|_0$$

$$\le c''[w_{|\alpha|-(n-\nu)}(x)]^{-1}\|u\|_0.$$

Multiplied by $w_{|\alpha|-(n-\nu)}(x)$, as $|D^\alpha v(x)|$ in (3.13), these two terms are estimated by $c_1\|u\|_0$.

Further, using (3.3) we estimate

$$\Big|\int_{B(x,r)} K_{\alpha-\beta}(x,y)D^\beta u(y)dy\Big| \le c\left\{\begin{matrix} r^n & , & \nu<0 \\ r^n(1+\log r), & & \nu=0 \\ r^{n-\nu} & , & \nu>0 \end{matrix}\right\} \sup_{y\in B(x,r)} |D^\beta u(y)|$$

and, due to (3.15), after the multiplication by $w_{|\alpha|-(n-\nu)}$ we obtain

$$w_{|\alpha|-(n-\nu)}(x)\Big|\int_{B(x,r)} K_{\alpha-\beta}(x,y)D^\beta u(y)dy\Big|$$

$$\le c'\left\{\begin{matrix} \rho(x)^n & , & \nu<0 \\ \rho(x)^n(1+|\log\rho(x)|), & & \nu=0 \\ \rho(x)^{n-\nu} & , & \nu>0 \end{matrix}\right\} \sup_{y\in B(x,r)} w_{|\alpha|-(n-\nu)}(y)|D^\beta u(y)|.$$

Summing up over β $(0\le\beta\le\alpha)$ we obtain a contribution which is not worse than $c_2\rho(x)^{\min\{1,n-\nu\}}\|u\|_{|\alpha|,\nu}$ in (3.13).

At last, using again (3.2) we estimate, for $1\le j\le|\alpha|-1$, $|\beta|\le j$,

$$\Big|\int_{S(x,r)}\Big[D_x^{\alpha-\alpha^{j+1}}K_{\alpha^j-\beta}(x,y)\Big]D^\beta u(y)\,\frac{x_{i_{j+1}}-y_{i_{j+1}}}{r}\,dS_y\Big|$$

$$\le cr^{n-1}\left\{\begin{matrix} 1 & , & \nu+|\alpha|-j-1<0 \\ 1+|\log r| & , & \nu+|\alpha|-j-1=0 \\ r^{-\nu-|\alpha|+j+1} & , & \nu+|\alpha|-j-1>0 \end{matrix}\right\} \sup_{y\in S(x,r)} |D^\beta u(y)|$$

and

$$w_{|\alpha|-(n-\nu)}(x)\Big|\int_{S(x,r)}\Big[D_x^{\alpha-\alpha^{j+1}}K_{\alpha^j-\beta}(x,y)\Big]D^\beta u(y)\,\frac{x_{i_{j+1}}-y_{i_{j+1}}}{r}\,dS_y\Big|$$

$$\le c'\left\{\begin{matrix} \rho(x)^{n-1} & , & \nu+|\alpha|-j-1<0 \\ \rho(x)^{n-1}(1+|\log\rho(x)|), & & \nu+|\alpha|-j-1=0 \\ \rho(x)^{n-\nu-|\alpha|+j} & , & \nu+|\alpha|-j-1>0 \end{matrix}\right\} \frac{w_{|\alpha|-(n-\nu)}(x)}{w_{|\beta|-(n-\nu)}(x)} \tag{3.16}$$

$$\times \sup_{y\in S(x,r)} w_{|\beta|-(n-\nu)}(y)|D^\beta u(y)|.$$

Some trouble causes only the case $\nu+|\alpha|-j-1>0$ (the third row). If thereby $n-\nu-|\alpha|+j\ge1$ then a suitable contribution to (3.13) is obtained again estimating $w_{|\alpha|-(n-\nu)}(x)/w_{|\beta|-(n-\nu)}(x)$ coarsely by a constant. Consider the case $n-\nu-|\alpha|+j<1$. Together with the inequality $j\geq 1$ we have then $|\alpha|>n-\nu$, therefore $w_{|\alpha|-(n-\nu)}(x)=\rho(x)^{|\alpha|-(n-\nu)}$ and

$$\rho(x)^{n-\nu-|\alpha|+j}\ \frac{w_{|\alpha|-(n-\nu)}(x)}{w_{|\beta|-(n-\nu)}(x)}$$

$$\leq \rho(x)^{n-\nu-|\alpha|+j}\left\{\begin{array}{ll}\rho(x)^{|\alpha|-(n-\nu)} & ,\ |\beta|<n-\nu\\ \rho(x)^{|\alpha|-(n-\nu)}(1+|\log\rho(x)|), & |\beta|=n-\nu\\ \rho(x)^{|\alpha|-|\beta|} & ,\ |\beta|>n-\nu\end{array}\right\}$$

$$=\left\{\begin{array}{ll}\rho(x)^{j} & ,\ |\beta|<n-\nu\\ \rho(x)^{j}(1+|\log\rho(x)|), & |\beta|=n-\nu\\ \rho(x)^{n-\nu+j-|\beta|} & ,\ |\beta|>n-\nu\end{array}\right\}\leq\left\{\begin{array}{ll}\rho(x) & ,\ |\beta|<n-\nu\\ \rho(x)(1+|\log\rho(x)|), & |\beta|=n-\nu\\ \rho(x)^{n-\nu} & ,\ |\beta|>n-\nu\end{array}\right\}.$$

Summing (3.16) up over β and j $(0\leq\beta\leq\alpha^{j},\ j=1,\ \dots\ ,|\alpha|-1)$ according to (3.14) we obtain a contribution

$$c_2\left\{\begin{array}{ll}\rho(x)^{\min\{1,n-\nu\}}\|u\|_{|\alpha|,\nu} & \text{if } \nu \text{ is a fraction}\\ \rho(x)(1+|\log\rho(x)|)\|u\|_{|\alpha|,\nu} & \text{if } \nu \text{ is an integer}\end{array}\right\}$$

This completes the proof of estimate (3.13).

It follows from (3.13) that $v\in C^{m,\nu}(G)$. Lemma 3.4 is proved.

Another consequence of (3.12) and (3.13) is that the operator T is bounded in $C^{m,\nu}(G)$.

3.5. Proof of Theorem 3.1. Let the kernel $K(x,y)$ satisfy condition (3.2) and let $f\in C^{m,\nu}(G)$. We have to prove that any solution $u\in L^{\infty}(G)$ of equation (3.1) belongs to $C^{m,\nu}(G)$. The idea that we follow is to replace equation (3.1) by an equation on a small open subset $\Omega\subseteq G$ such that the operator $I-T_{\Omega}$ is invertible in both spaces $L^{\infty}(\Omega)$ and $C^{m,\nu}(\Omega)$. The operator T_{Ω} is defined via the formula

$$(T_{\Omega}u)(x)=\int_{\Omega}K(x,y)u(y)dy,\quad x\in\Omega.$$

Let us define (cf. (2.11))

$$w_{\lambda,\Omega}(x)=\left\{\begin{array}{ll}1 & \lambda<0\\ [1+|\log\rho_{\Omega}(x)|]^{-1}, & \lambda=0\\ [\rho_{\Omega}(x)]^{\lambda} & ,\ \lambda>0\end{array}\right\},\ x\in\Omega,$$

where $\rho_\Omega(x)$ is the distance from $x \in \Omega$ to $\partial\Omega$. Introduce the norm in $C^{m,\nu}(\Omega)$ in a similar way as in Section 2.5:

$$\|u\|_{m,\nu,\Omega} = \sum_{|\alpha| \le m} \sup_{x \in \Omega} (w_{|\alpha|-(n-\nu),\Omega}(x)|D^\alpha u(x)|).$$

According to Lemma 3.4, the operator T_Ω maps $C^{m,\nu}(\Omega)$ into $C^{m,\nu}(\Omega)$, and, for any $u \in C^{m,\nu}(\Omega)$,

$$\|T_\Omega u\|_{0,\Omega} \le c_{0,\Omega}\|u\|_{0,\Omega}, \qquad c_{0,\Omega} \to 0 \ \text{ as } \ \operatorname{diam}\Omega \to 0,$$

$$w_{|\alpha|-(n-\nu),\Omega}(x)|D^\alpha(T_\Omega u)(x)|$$

$$\le \left\{ \begin{array}{ll} c_1\|u\|_{0,\Omega} + c_2\rho_\Omega(x)^{\min\{1,n-\nu\}}\|u\|_{|\alpha|,\nu,\Omega} & , \ \nu \notin \mathbb{Z}, \\ c_1\|u\|_{0,\Omega} + c_2\rho_\Omega(x)(1+|\log\rho_\Omega(x)|)\,\|u\|_{|\alpha|,\nu,\Omega}, & \nu \in \mathbb{Z} \end{array} \right\}, \quad x \in \Omega,$$

where $\|u\|_{0,\Omega} = \sup_{x\in\Omega}|u(x)|$. Since $\rho_\Omega(x) < \operatorname{diam}\Omega$, there exist a sufficiently small $d_0 > 0$ and a sufficiently great $M \ge 1$ such that, for any open set $\Omega \subset G$ with $\operatorname{diam}\Omega \le d_0$ and any $u \in C^{m,\nu}(\Omega)$,

$$\|T_\Omega u\|_{0,\Omega} \le \frac{1}{4}\|u\|_{0,\Omega}, \tag{3.17}$$

$$\sum_{1\le|\alpha|\le m} \sup_{x\in\Omega}(w_{|\alpha|-(n-\nu),\Omega}(x)|D^\alpha(T_\Omega u)(x)|)$$
$$\le M\|u\|_{0,\Omega} + \frac{1}{2}\sum_{1\le|\alpha|\le m} \sup_{x\in\Omega}(w_{|\alpha|-(n-\nu),\Omega}(x)|D^\alpha u(x)|). \tag{3.18}$$

A consequence is that $I - T_\Omega$ is invertiable in $L^\infty(\Omega)$ and $C^{m,\nu}(\Omega)$ whereby

$$\|(I - T_\Omega)^{-1}\|_{\mathcal{L}(C^{m,\nu}(\Omega),C^{m,\nu}(\Omega))} \le 8M. \tag{3.19}$$

Indeed, the invertibility in $L^\infty(\Omega)$ follows immediately from (3.17). Let us prove the invertibility in $C^{m,\nu}(\Omega)$ and estimate (3.19). Introduce in $C^{m,\nu}(\Omega)$ a provisional new norm

$$\|u\|'_{m,\nu,\Omega} = 4M\|u\|_{0,\Omega} + \sum_{1\le|\alpha|\le m} \sup_{x\in\Omega}(w_{|\alpha|-(n-\nu),\Omega}(x)|D^\alpha u(x)|)$$

which is equivalent with the old one:

$$\|u\|_{m,\nu,\Omega} \le \|u\|'_{m,\nu,\Omega} \le 4M\|u\|_{m,\nu,\Omega}.$$

Using (3.17) and (3.18) we find

$$\|T_\Omega u\|'_{m,\nu,\Omega} = 4M\,\|T_\Omega u\|_{0,\Omega} + \sum_{1\le|\alpha|\le m} \sup_{x\in\Omega}(w_{|\alpha|-(n-\nu),\Omega}(x)\,|D^\alpha(T_\Omega u)(x)|)$$

$$\le 2M\,\|u\|_{0,\Omega} + \frac{1}{2}\sum_{1\le|\alpha|\le m} \sup_{x\in\Omega}(w_{|\alpha|-(n-\nu),\Omega}(x)\,|D^\alpha u(x)|)$$

$$= \frac{1}{2}\,\|u\|'_{m,\nu,\Omega}, \qquad u\in C^{m,\nu}(\Omega).$$

Thus, with respect to the provisional norm, $\|T_\Omega\|_{\mathscr{L}(C^{m,\nu}(\Omega),C^{m,\nu}(\Omega))} \le 1/2$ and $\|(I-T_\Omega)^{-1}\|_{\mathscr{L}(C^{m,\nu}(\Omega),C^{m,\nu}(\Omega))} \le 2$. This yields estimate (3.19) with respect to the norm $\|u\|_{m,\nu,\Omega}$. From now we work only with the norm $\|u\|_{m,\nu,\Omega}$.

Let $u_0\in L^\infty(G)$ be a solution to equation (3.1). We have to prove that $u_0\in C^{m,\nu}(G)$. Fix an arbitrary point $\bar{x}\in G$ and introduce the set $\Omega = B(\bar{x},d_0/2)\cap G$. Thus, $\operatorname{diam}\Omega\le d_0$, and estimate (3.19) holds. Notice that the restriction of u_0 to Ω satisfies the equation

$$u(x) = \int_\Omega K(x,y)u(y)\,dy + f_\Omega(x), \quad x\in\Omega, \tag{3.20}$$

where

$$f_\Omega(x) = f(x) + \int_{G\setminus\Omega} K(x,y)u_0(y)\,dy, \quad x\in\Omega.$$

An important observation is that $f_\Omega\in C^{m,\nu}(\Omega)$. Indeed, for $x\in\Omega$ and $y\in G\setminus\Omega$ we have $|x-y|\ge\rho_\Omega(x)>0$, and we may differentiate the function $\int_{G\setminus\Omega}K(x,y)u_0(y)\,dy$ under the integral sign. The result of differentation

$$D^\alpha \int_{G\setminus\Omega} K(x,y)u_0(y)\,dy = \int_{G\setminus\Omega} D_x^\alpha K(x,y)u_0(y)\,dy, \quad |\alpha|\le m$$

is a continuous function on Ω. Further, using (3.4) we estimate

$$\left|D^\alpha \int_{G\setminus\Omega} K(x,y)u_0(y)\,dy\right| \le \int_{G\setminus B(x,\rho_\Omega(x))} |D_x^\alpha K(x,y)|\,dy\,\|u\|_0$$

$$\le b\begin{Bmatrix} 1, & |\alpha|<n-\nu \\ 1+|\log\rho_\Omega(x)|, & |\alpha|=n-\nu \\ \rho_\Omega(x)^{n-\nu-|\alpha|}, & |\alpha|>n-\nu \end{Bmatrix}\|u\|_0, \quad x\in\Omega,\ |\alpha|\le m.$$

We see that $\int_{G\setminus\Omega}K(x,y)u_0(y)\,dy$ belongs to $C^{m,\nu}(\Omega)$ and

$$\|f_\Omega\|_{m,\nu,\Omega} \le \|f\|_{m,\nu} + c\,\|u_0\|_0, \quad c = b\sum_{|\alpha|\le m} 1 = \text{const}. \tag{3.21}$$

Thus, equation (3.20) is uniquely solvable in $C^{m,\nu}(\Omega)$. But equation (3.20) is uniquely solvable in $L^\infty(\Omega)$, too, and we know the solution — it is u_0, the restriction to Ω of the solution of (3.1) under consideration. Since

$C^{m,\nu}(\Omega)\subset L^\infty(\Omega)$, the solutions in $C^{m,\nu}(\Omega)$ and $L^\infty(\Omega)$ coincide. In other words, $u_0\in C^{m,\nu}(\Omega)$. Due to (3.19) and (3.21),

$$\|u_0\|_{m,\nu,\Omega} \le 8M(\|f\|_{m,\nu} + c\|u_0\|_0)$$

Especially, for the point $\bar{x}\in\Omega = G\cap B(\bar{x},d_0/2)$ we have

$$|D^\alpha u_0(\bar{x})| \le 8M \begin{Bmatrix} 1 & , & |\alpha|<n-\nu \\ 1+|\log\rho_\Omega(\bar{x})| & , & |\alpha|=n-\nu \\ [\rho_\Omega(\bar{x})]^{n-\nu-|\alpha|} & , & |\alpha|>n-\nu \end{Bmatrix} (\|f\|_{m,\nu} + c\|u_0\|_0), \quad |\alpha|\le m.$$

Since $\bar{x}\in G$ is arbitrary and, for $\Omega = G\cap B(\bar{x},d_0/2)$, $\rho_\Omega(\bar{x}) = \min\{\rho(\bar{x}),d_0/2\}$,

$$(d_0/\operatorname{diam} G)\,\rho(\bar{x}) \le \rho_\Omega(\bar{x}) \le \rho(\bar{x}),$$

we obtain that $u_0\in C^{m,\nu}(G)$, q.e.d.

To finish with the proof of Theorem 3.1, we remark that any solution $u_0\in L(G)$ of equation (3.1) belongs to $L^\infty(G)$. Here the improving property of the weakly singular integral operator T can be used: there are numbers $1=p_1<p_2<\ldots<p_l<\infty$ such that T maps $L^{p_i}(G)$ into $L^{p_{i+1}}(G)$, $i=1, \ldots, l-1$, and $L^{p_l}(G)$ into $L^\infty(G)$. We successevely see from the equality $u_0 = Tu_0+f$ that $u_0\in L^{p_i}(G)$, $i=1, \ldots, l$, and finally $u_0\in L^\infty(G)$. Another possibility to establish the inclusion $u_0\in L^\infty(G)$ is to use the following general result putting $E=L^\infty(G)$, $E_1=L^1(G)$.

Lemma 3.5. Let E and E_1 be Banach spaces whereby $E\subset E_1$ continuously and densely. Let $T\in\mathcal{L}(E,E)$ and $T_1\in\mathcal{L}(E_1,E_1)$ be compact operators whereby $T\subset T_1$, i.e. T is the restriction of T_1 to E. If the equation $u=T_1u+f$ with $f\in E$ has a solution $u\in E_1$ then $u\in E$ and $\|u\|_E\le c_1\|u\|_{E_1}+c\|f\|_E$ where the constants c_1 and c are independent of u and f.

This Lemma is an elementary consequence of the Riesz-Schauder theory about compact operators. The Lemma remains true if $T\in\mathcal{L}(E,E)$ and $T_1\in\mathcal{L}(E_1,E_1)$, being possibly non-compact, have compact powers $T^k\in\mathcal{L}(E,E)$ and $T_1^k\in\mathcal{L}(E_1,E_1)$ where k is an integer, $k\ge1$. This remark enables to construct an alternative proof of Theorem 3.1 putting $E=C^{m,\nu}(G)$, $E_1=L^1(G)$ and showing that $T^2\in\mathcal{L}(C^{m,\nu}(G),C^{m,\nu}(G))$ is compact for the integral operator T of equation (3.1). This proof idea was realized in Vainikko (1991); the proof of the compactness of $T^2\in\mathcal{L}(C^{m,\nu}(G),C^{m,\nu}(G))$ is based on Lemma 3.4.

First proof of a result like Theorem 3.1 seems to belong to J. Pitkäranta (1979,1980). He followed the simplest idea to examine the improving properties of T from L(G) up to a space of the type $C^{m,\nu}(G)$. This way is rather tecnical.

3.6. Tangential differentiation of weakly singular integral. For a vector field $a(x)$ of the class $[C^m(\overline{G})]^n$, let us denote

$$K_a(x,y) = \sum_{i=1}^{n}\Big[a_i(x)\frac{\partial K(x,y)}{\partial x_i} + a_i(y)\frac{\partial K(x,y)}{\partial y_i}\Big] + K(x,y)\sum_{i=1}^{n}\frac{\partial a_i(y)}{\partial y_i}$$

$$= (\mathcal{L}_{a,x} + \mathcal{L}_{a,y})K(x,y) + K(x,y)\operatorname{div} a(y), \quad x,y \in G. \tag{3.22}$$

It follows from (3.2) that $K_a(x,y)$ also satisfies (3.2) with the same ν (but of course with $m-1$ instead of m). Multiplying (3.11) by $a_i(x)$, adding the equalities for $i=1, \dots, n$, and grouping, we get that

$$\mathcal{L}_a \int_\Omega K(x,y)u(y)dy = \int_\Omega K_a(x,y)u(y)dy$$

$$-\int_\Omega \sum_{i=1}^{n}[a_i(y)-a_i(x)]\frac{\partial K(x,y)}{\partial y_i}u(y)dy - \int_\Omega K(x,y)\sum_{i=1}^{n}\frac{\partial a_i(y)}{\partial y_i}u(y)dy$$

$$+\int_\Omega K(x,y)\sum_{i=1}^{n}a_i(x)\frac{\partial u(y)}{\partial y_i}dy + \int_{\partial\Omega}K(x,y)u(y)\sum_{i=1}^{n}a_i(x)\omega_i(y)dS_y.$$

We integrate the second term on the right by parts (this is easily justified with the help of (3.2)), getting

$$\mathcal{L}_a \int_\Omega K(x,y)u(y)dy = \int_\Omega K_a(x,y)u(y)dy + \int_\Omega K(x,y)\mathcal{L}_a u(y)dy$$

$$+\int_{\partial\Omega}K(x,y)\sum_{i=1}^{n}a_i(x)\omega_i(y)u(y)dS_y, \quad x\in\Omega, \tag{3.23}$$

which is valid under the conditions of Lemma 3.3, e.g. for $u \in C^1(\overline{\Omega})$. However, the requirements on u in Lemma 3.3 are too restrictive for our subsequent goals, and we weaken them.

__Lemma__ **3.6.** Suppose that ∂G is piecewise smooth and $a(x)$ is vector field of class $[C^m(\overline{G})]^n$. Assume that condition (3.2) holds. Let $x^0 \in G$ be a particular point, and let $0<r<\rho_a(x^0)$. Then the following formula holds for a continuously differentiable function u on $\bar{B}(x^0,r)\cap G$ which is bounded on $\bar{B}(x^0,r)\cap G$ together with $\mathcal{L}_a u$:

$$\mathcal{L}_a \int_{B(x^0,r)\cap G} K(x,y)u(y)dy$$

$$= \int_{B(x^0,r)\cap G} K_a(x,y)u(y)dy + \int_{B(x^0,r)\cap G} K(x,y)\mathcal{L}_a u(y)dy +$$

$$+\int\limits_{S(x^0,r)\cap G} K(x,y)\sum_{i=1}^{n} a_i(x)\omega_i(y)u(y)\,dS_y, \quad x\in B(x^0,r)\cap G, \tag{3.24}$$

where $\omega(y)=(x^0-y)/r$ is the inner normal to $S(x^0,r)$ at the point $y\in S(x^0,r)\cap G$.

Proof. In general, the ball $B(x^0,r)$ contains some piecewise smooth hypersurfaces making up ∂G, and field a is tangent to them (see the definition of $\rho_a(x)$ in Section 2.6) We split off from $\Omega=B(x^0,r)\cap G$ a small neighborhood $\mathcal{N}_\varepsilon$ (see Figure 3.2) of the set $\partial G\cap B(x^0,r)$ such that the volume measure of $\mathcal{N}_\varepsilon$ and the surface measure of $S(x^0,r)\cap\overline{\mathcal{N}}_\varepsilon$ tend to zero as $\varepsilon\to 0$, and such that the new part of the boundary of the set $\Omega_\varepsilon=\Omega\setminus\mathcal{N}_\varepsilon$ (i.e., $\partial\Omega_\varepsilon$, excluding $S(x^0,r)$) maintains the property of being tangent to a. It follows from the continuous differentiability of u in $\bar{B}(x^0,r)\cap G$ (see the formulation of the Lemma) that $u\in C^1(\bar{\Omega}_\varepsilon)$ and that (3.23) is applicable on the set Ω_ε. Further, $\sum_{i=1}^{n}a_i(y)\omega_i(y)=0$ on the part of $\partial\Omega_\varepsilon$ tangent to a, so that the surface integral in (3.23) stays only over the set $[S(x^0,r)\cap G]\setminus[S(x^0,r)\cap\overline{\mathcal{N}}_\varepsilon]$. We get (3.24) in the limit as $\varepsilon\to 0$. Lemma 3.6 is proved.

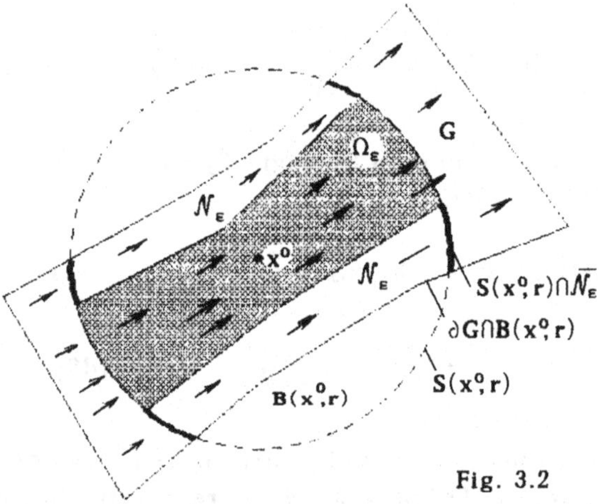

Fig. 3.2

We get the following result from Lemma 3.5 by induction.

Lemma 3.7. Assume the condition of Lemma 3.6. Then a function u which is m times continuously differentiable in $\bar{B}(x^0,r)\cap G$ and is bounded on this set together with $\mathcal{L}_a u, \ldots, \mathcal{L}_a^m u$ satisfies

$$\mathcal{L}_a^k\int\limits_{B(x^0,r)\cap G}K(x,y)u(y)\,dy=\sum_{j=0}^{k}\binom{k}{j}\int\limits_{B(x^0,r)\cap G}K_a^{(k-j)}(x,y)\mathcal{L}_a^j u(y)\,dy$$

$$+\sum_{j=0}^{k-1}\sum_{l=0}^{j}\binom{j}{l}\int\limits_{S(x^0,r)\cap G}\mathcal{L}_{a,x}^{k-1-j}K_a^{(j-l)}(x,y)\sum_{i=1}^{n}a_i(y)\omega_i(y)\mathcal{L}_a^l u(y)\,dy, \tag{3.25}$$

$$x\in B(x^0,r),\quad k=1,\ldots,m,$$

where $\mathscr{L}_a^0 u = u$, $K^{(0)}(x,y) = K(x,y)$, $K_a^{(1)}(x,y) = K_a(x,y)$ (see (3.22)) and

$$K_a^{(j+1)}(x,y) = \sum_{i=1}^{n} \Big[a_i(x) \frac{\partial K_a^{(j)}(x,y)}{\partial x_i} + a_i(y) \frac{\partial K_a^{(j)}(x,y)}{\partial y_i} \Big] + K_a^{(j)}(x,y) \sum_{i=1}^{n} \frac{\partial a_i(y)}{\partial y_i}. \tag{3.26}$$

It follows by induction from (3.2) that

$$|\mathscr{L}_{a,x}^k K_a^{(j)}(x,y)| \le b' \left\{ \begin{array}{ll} 1 & , \quad \nu + k < 0 \\ 1 + |\log|x-y|| , & \nu + k = 0 \\ |x-y|^{-\nu-k} & , \quad \nu + k > 0 \end{array} \right\}, \quad k + j \le m. \tag{3.27}$$

3.7. Estimates of the tangential derivatives of Tu. The following result is similar to Lemma 3.4.

Lemma 3.8. Suppose that ∂G is piecewise smooth, $a(x)$ is a vector field of class $[C^m(\overline{G})]^n$ and condition (3.2) for the kernel $K(x,y)$ holds. Then, for any $u \in C_a^{m,\nu}(G)$, the function

$$v(x) = (Tu)(x) = \int_G K(x,y) u(y) dy$$

also belongs to $C_a^{m,\nu}(G)$. In addition to (3.12) and (3.13), the following pointwise estimate hold for $x \in G$, $k = 1, \dots, m$:

$$w_{k-(n-\nu),a}(x) |\mathscr{L}_a^k v(x)|$$

$$\le \left\{ \begin{array}{ll} c_1 \|u\|_0 + c_2 \rho_a(x)^{\min\{1,n-\nu\}} \|u\|_{k,\nu,a} , & \nu \notin \mathbb{Z}, \\ c_1 \|u\|_0 + c_2 \rho_a(x)(1 + |\log \rho_a(x)|) \|u\|_{k,\nu,a} , & \nu \in \mathbb{Z} \end{array} \right\}. \tag{3.28}$$

Proof. Consider an arbitrary point $x^0 \in G$. Let $0 < r < \rho_a(x^0)$. We have

$$v(x) = \int_{G \setminus B(x^0,r)} K(x,y) u(y) dy + \int_{B(x^0,r) \cap G} K(x,y) u(y) dy .$$

For $x \in B(x^0, r) \cap G$, the first integral may be differentiated under the integral sign; for the second integral we have formula (3.25). After the applying $\mathscr{L}_a^k$ we put $x = x^0$ (maintaining the designation x istead of x^0). Thus we obtain the formula (cf.(3.14))

$$\mathscr{L}_a^k v(x) = \int_{G \setminus B(x,r)} \mathscr{L}_{a,x}^k K(x,y) u(y) dy$$

$$+ \sum_{j=0}^{k} \binom{k}{j} \int_{B(x,r) \cap G} K_a^{(k-j)}(x,y) \mathscr{L}_a^j u(y) dy +$$

$$+ \int_{S(x,r)\cap G} \mathcal{L}_{a,x}^{k-j} K(x,y) \sum_{i=1}^{n} a_i(y)\omega_i(y)u(y)dy \tag{3.29}$$
$$+ \sum_{j=1}^{k-1} \sum_{l=0}^{j} \binom{j}{l} \int_{S(x,r)\cap G} \mathcal{L}_{a,x}^{k-1-j} K_a^{(j-l)}(x,y) \sum_{i=1}^{n} a_i(y)\omega_i(y)\mathcal{L}_a^{l} u(y)dy,$$

which holds for $x\in G$, $0<r<\rho_a(x)$, $k=1,\dots,m$. Again, as in (3.14), we have raised the surface integral with $j=0$ separately. Putting $r=\rho_a(x)/2$ and estimating the terms of expansion (3.29) by the help of (3.27) in a very similar way as in the proof of Lemma 3.4 we obtain (3.28); a detailed estimation is proposed to the reader as an exercise. It follows from (3.12), (3.13) and (3.28) that $v\in C_a^{m,\nu}(G)$. Lemma 3.8 is proved.

3.8. Proof of Theorem 3.2 is similar to the proof of Theorem 3.1 given in Section 3.5, therefore we present here only the outlines leaving the details to the reader.

Let the kernel $K(x,y)$ satisfy (3.2) and let $f\in C_a^{m,\nu}(G)$. We know (see Theorem 3.1) that any solution $u_0\in L^\infty(G)$ (or even $u_0\in L(G)$) belongs to $C^{m,\nu}(G)$; we have to prove that u_0 belongs $C_a^{m,\nu}(G)$. Fix an arbitrary point $\bar{x}\in G$ and introduce the set $\Omega=B(\bar{x},d_0/2)\cap G$ where d_0 is a sufficiently small positive constant so that, for any $u\in C_a^{m,\nu}(G)$, inequalities (3.17),(3.18) and the supplementary inequality

$$\sum_{k=1}^{m} \sup_{x\in\Omega} \left(w_{k-(n-\nu),a,\Omega}(x)\,|\mathcal{L}_a^k(Tu)(x)|\right) \tag{3.30}$$
$$\le M\|u\|_{0,\Omega} + \frac{1}{2}\sum_{k=1}^{m} \sup_{x\in\Omega} \left(w_{k-(n-\nu),a,\Omega}(x)\,|\mathcal{L}_a^k u(x)|\right)$$

hold , where

$$w_{\lambda,a,\Omega}(x) = \begin{cases} 1 & , \ \lambda<0 \\ [1+|\log\rho_{a,\Omega}(x)|]^{-1}, & \lambda=0 \\ [\rho_{a,\Omega}(x)]^\lambda & , \ \lambda>0 \end{cases}, \quad x\in\Omega,$$

$$\rho_{a,\Omega}(x) = \inf_{y\in\partial\Omega\setminus\Gamma_a} |x-y|, \quad x\in\Omega.$$

Inequality (3.30) is a consequence of Lemma 3.8 applied to $\Omega=B(\bar{x},d_0/2)\cap G$ instead of whole G. Note that the constant M in (3.30) as well d_0 are independent of the particular choice of the point $\bar{x}\in G$. A corollary of (3.17), (3.18) and (3.30) is that the operator $I-T_\Omega$ is invertible in $C^{m,\nu}(\Omega)$ and $C_a^{m,\nu}(\Omega)$ whereby

$$\|(I-T_\Omega)^{-1}\|_{\mathcal{L}(C_a^{m,\nu}(\Omega),C_a^{m,\nu}(\Omega))} \le 16M.$$

Representing equation (3.1) in the form (3.20) it is easy to check that $f_\Omega \in C_a^{m,\nu}(\Omega)$ and

$$\|f_\Omega\|_{m,\nu,a,\Omega} \le \|f\|_{m,\nu,a} + c\|u_0\|_0.$$

Equation (3.20) is uniquely solvable in $C^{m,\nu}(\Omega)$ and $C_a^{m,\nu}(\Omega)$, and we know the solution in $C^{m,\nu}(\Omega)$ — it is u_0. Therefore, $u_0 \in C_a^{m,\nu}(\Omega)$,

$$\|u_0\|_{m,\nu,a,\Omega} \le 16M(\|f\|_{m,\nu,a} + c\|u_0\|_0).$$

Especially, for our point $\bar{x} \in \Omega = G \cap B(\bar{x}, d_0/2)$, we have

$$|\mathcal{L}_a^k u_0(\bar{x})| \le 16M \begin{Bmatrix} 1 & , & k < n-\nu \\ 1 + |\log \rho_{a,\Omega}(\bar{x})| & , & k = n-\nu \\ [\rho_{a,\Omega}(\bar{x})]^{n-\nu-k} & , & k > n-\nu \end{Bmatrix} (\|f\|_{m,\nu,a} + c\|u_0\|_0), \quad k = 1, \dots, m.$$

Since $\bar{x} \in G$ is arbitrary and, for $\Omega = G \cap B(\bar{x}, d_0/2)$, $\rho_{a,\Omega}(\bar{x}) = \min\{\rho_a(\bar{x}), d_0/2\}$,

$$(d_0/2\,\mathrm{diam}\,G)\rho_a(\bar{x}) \le \rho_{a,\Omega}(\bar{x}) \le \rho_a(\bar{x}),$$

we obtain that $u_0 \in C_a^{m,\nu}(G)$. This completes the proof of Theorem 3.2.

3.9. Sharpness of Theorem 3.1 (general analysis). Now we try to select main singular parts of the derivatives of a solution u to equation (3.1) and in this way to show that, under some conditions, Theorem 3.1 provides a sharp characterization of singularities. We assume here that $f \in C^m(\bar{G})$ and $G \subset \mathbb{R}^n$ is a sufficiently regular bounded region such that

$$d_G(x,y) \le c_G|x-y|, \quad \forall x,y \in G, \tag{3.31}$$

where c_G is a constant. Then due to Lemma 2.3, a solution u of equation (3.1) is Hölder continuous on $\bar{G}$. More precisely, for any $x^1, x^2 \in G$,

$$|u(x^1) - u(x^2)| \le \mathrm{const} \begin{Bmatrix} |x^1 - x^2| & , & \nu < n-1 \\ |x^1 - x^2|(1 + \log|x^1 - x^2|), & & \nu = n-1 \\ |x^1 - x^2|^{n-\nu} & , & \nu > n-1 \end{Bmatrix},$$

and this inequality can be extended to $\bar{G}$. Further, it follows from (3.2) that the derivatives of $K(x,y)$ up to the order m-1 are continuous on $(\bar{G} \times \bar{G}) \setminus \{x=y\}$.

__Lemma__ **3.9.** Assume $G \subset \mathbb{R}^n$ is open bounded and satisfies (3.31), ∂G is piecewise smooth, $f \in C^m(\bar{G})$ and K satisfies (3.2). Then for any solution $u \in C^{m,\nu}(G)$ of equation (3.1) and any multi-index α with $|\alpha| \ge n-\nu$, $1 \le |\alpha| \le m$, representing it as $\alpha = e^l + \alpha'$, $1 \le l \le n$, we have

$$w_{|\alpha|-(n-\nu)}(x) \left| D^\alpha u(x) - \int_{\partial G} D_x^{\alpha'} K(x,y) u(y) \omega_l(y) dS_y \right| \to 0 \text{ as } \rho(x) \to 0 \tag{3.32}$$

where $\omega(y)=(\omega_1(y),\dots,\omega_n(y))$ is the unit inner normal to ∂G.

Proof. Differentiating the equality $u=f+Tu$ we obtain $D^\alpha u = D^\alpha f + D^\alpha Tu$; the term $D^\alpha v = D^\alpha Tu$ is given by formula (3.14) where $\overline{B}(x,r)\subset G$, $i_1=i$, $\alpha-\alpha^1=\alpha'$. We transform the first integral in (3.14) representing

$$D_x^\alpha = \frac{\partial}{\partial x_i} D_x^{\alpha'} = D_x^{\alpha'}\Big(\frac{\partial}{\partial x_i}+\frac{\partial}{\partial y_i}\Big) - \frac{\partial}{\partial y_i} D_x^{\alpha'}$$

and taking into account that

$$-\int_{G\setminus B(x,r)} \frac{\partial}{\partial y_i}\Big[D_x^{\alpha'}K(x,y)u(y)\Big]dy$$

$$= \int_{\partial G} D_x^{\alpha'}K(x,y)u(y)\omega_i(y)dS_y - \int_{S(x,r)} \Big[D_x^{\alpha'}K(x,y)\Big]u(y)\,\frac{x_i-y_i}{r}\,dS_y .$$

As the result we obtain the formula

$$D^\alpha u(x) = \int_{\partial G} D_x^{\alpha'}K(x,y)u(y)\omega_i(y)dS_y + D^\alpha f(x)$$

$$+ \int_{G\setminus B(x,r)} D_x^{\alpha'}\Big(\frac{\partial K(x,y)}{\partial x_i}+\frac{\partial K(x,y)}{\partial y_i}\Big)u(y)dy + \int_{G\setminus B(x,r)} D_x^{\alpha'}K(x,y)\,\frac{\partial u(y)}{\partial y_i}\,dy$$

$$+ \sum_{0\le\beta\le\alpha}\binom{\alpha}{\beta}\int_{B(x,r)} K_{\alpha-\beta}(x,y)D^\beta u(y)dy \tag{3.33}$$

$$+\sum_{j=1}^{|\alpha|-1}\ \sum_{0\le\beta\le\alpha^j}\binom{\alpha^j}{\beta}\int_{S(x,r)}\Big[D_x^{\alpha-\alpha^{j+1}}K_{\alpha^j-\beta}(x,y)\Big]D^\beta u(y)\,\frac{x_{i_{j+1}}-y_{i_{j+1}}}{r}\,dS_y .$$

To prove (3.32), it sufficies to notice that

$$w_{|\alpha|-(n-\nu)}(x)\,|D^\alpha f(x)| \le c\,w_{|\alpha|-(n-\nu)}(x) \to 0 \quad \text{as} \quad \rho(x)\to 0$$

and to prove that, with $r=\rho(x)/2$,

$$w_{|\alpha|-(n-\nu)}(x)\Big|\int_{G\setminus B(x,r)} D_x^{\alpha'}K(x,y)\,\frac{\partial u(y)}{\partial y_i}\,dy\Big|\to 0 \quad \text{as} \quad \rho(x)\to 0. \tag{3.34}$$

Compared with the proof of Lemma 3.4, this is the only new type term; for other integrals over $G\setminus B(x,r)$, $B(x,r)$ and $S(x,r)$ with $r=\rho(x)/2$ we have suitable estimates given in the proof of Lemma 3.4.

If $\nu\le n-1$ then $\partial u/\partial y_i\in L^p(G)$ with any $p<\infty$, and (3.34) can be obtained using the Hölder inequality:

$$\Big|\int_{G\setminus B(x,r)} D_x^{\alpha'}K(x,y)\,\frac{\partial u(y)}{\partial y_i}\,dy\Big| \le \Big(\int_{G\setminus B(x,r)} |D_x^{\alpha'}K(x,y)|^q\,dy\Big)^{1/q}\,\|\partial u/\partial y_i\|_{L^p(\Omega)},$$

$$p^{-1}+q^{-1}=1.$$

In the most heavy case where $\nu+|\alpha'|\geq n$ we have (cf. the proof of (3.4))

$$\Big(\int_{G\setminus B(x,r)} |D_x^{\alpha'}K(x,y)|^q\,dy\Big)^{1/q} \leq \Big(b\int_{r<|x-y|<d} |x-y|^{(-\nu-|\alpha'|)q}dy\Big)^{1/q}$$

$$\leq c(r^{n-(\nu+|\alpha'|)q})^{1/q} = cr^{(n/q)-\nu-|\alpha|+1} = c'\rho(x)^{(n/q)-n+1}\rho(x)^{n-\nu-|\alpha|},$$

therefore

$$w_{|\alpha|-(n-\nu)}(x)\Big|\int_{G\setminus B(x,r)} D_x^{\alpha'}K(x,y)\,\frac{\partial u(y)}{\partial y_i}\,dy\Big| \leq c\,\rho(x)^{(n/q)-n+1}.$$

Taking $p>n$, the exponent $(n/q)-n+1 = -(n/p)+1$ will be positive, and (3.34) follows. The reader can arrive to the same conclusion in the case $\nu+|\alpha'|<n$.

It remains to examine the case $n-1<\nu<n$. Then

$$|\partial u(y)/\partial y_i| \leq c\,\rho(y)^{n-\nu-1}, \quad \partial u/\partial y_i \in L(G).$$

We devide the integral in (3.34) into two parts. First, due to (3.4),

$$\Big|\int_{\{y\in G\setminus B(x,r):\rho(y)\geq r\}} D_x^{\alpha'}K(x,y)\,\frac{\partial u(y)}{\partial y_i}\,dy\Big| \leq c\int_{G\setminus B(x,r)} |D_x^{\alpha'}K(x,y)|dy\; r^{n-\nu-1}$$

$$\leq c'r^{n-\nu-|\alpha|+1}r^{n-\nu-1} \leq c''\rho(x)^{n-\nu}\rho(x)^{n-\nu-|\alpha|}$$

if $|\alpha'|\geq 1$ (implying $\nu+|\alpha'|>n$); for $\alpha'=0$ we estimate

$$\Big|\int_{\{y\in G\setminus B(x,r):\rho(y)\geq r\}} K(x,y)\,\frac{\partial u(y)}{\partial y_i}\,dy\Big| \leq c\int_{\{y\in G:\,\rho(y)\geq r\}} |x-y|^{-\nu}\rho(y)^{n-\nu-1}dy$$

$$\leq c\int_G |x-y|^{-\nu}\rho(y)^{-\Theta}dy\, r^{n-\nu-1+\Theta} \leq c'\rho(x)^{\Theta}\rho(x)^{n-\nu-1}$$

where $0<\Theta<\min\{1-(n-\nu),\,(n-\nu)/n\}$. The result supports (3.34) in both cases. Secondly, according to (3.2),

$$\Big|\int_{\{y\in G\setminus B(x,r):\rho(y)<r\}} D_x^{\alpha'}K(x,y)\,\frac{\partial u(y)}{\partial y_i}\,dy\Big|$$

$$\leq bc\int_{\{y\in G\setminus B(x,r):\rho(y)<r\}} |x-y|^{-\nu-|\alpha|+1}\rho(y)^{n-\nu-1}dy$$

$$\leq c'\Big(1+\int_{\{y\in G:\,r<|x-y|<r_0,\,\rho(y)<r\}} |x-y|^{-\nu-|\alpha|+1}\rho(y)^{n-\nu-1}dy\Big)$$

where we fixed a $r_0>0$. Using a rectifying transformation (see Figure 3.3) the last integral can be estimated by the integral

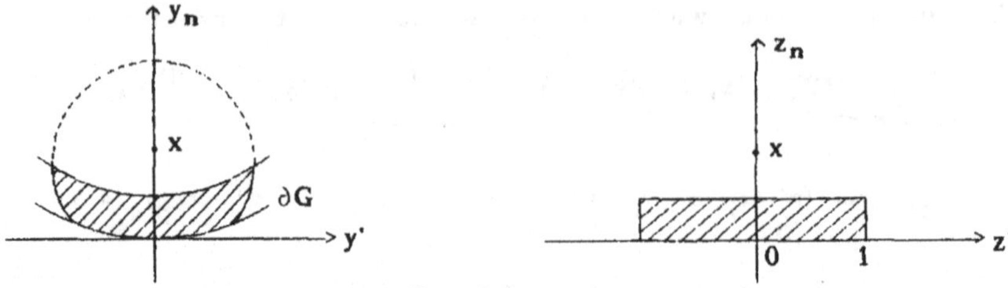

Fig. 3.3

$$\int_{\{z'\in\mathbb{R}^{n-1}:\,|z'|<1\}} (|z'|+r)^{-\nu-|\alpha|+1}dz' \int_0^r z_n^{n-\nu-1}dz_n$$

$$= \sigma_{n-1}\int_0^1 (\rho+r)^{-\nu-|\alpha|+1}\rho^{n-2}d\rho\ \frac{r^{n-\nu}}{n-\nu} \le c r^{n-\nu}\int_0^1 (\rho+r)^{n-\nu-|\alpha|-1}\,d\rho$$

$$\le c' r^{n-\nu} r^{n-\nu-|\alpha|} \le c'' \rho(x)^{n-\nu}\rho(x)^{n-\nu-|\alpha|}$$

and this supports (3.34) again.

The proof of Lemma 3.9 is finished.

Using the rectifying transformation again it is easy to see that

$$w_{|\alpha|-(n-\nu)}(x)\int_{\partial G}|D_x^{\alpha'}K(x,y)|\,dS_y \le \text{const}\ (x\in G)$$

and, for a smoothness point x^0 of ∂G,

$$w_{|\alpha|-(n-\nu)}(x)\left|\int_{\partial G} D_x^{\alpha'}K(x,y)u(y)\omega_i(y)dS_y - u(x^0)\omega_i(x^0)\int_{\partial G} D_x^{\alpha'}K(x,y)dS_y\right| \to 0$$

as $x\to x^0$. The last relation together with (3.32) implies the following result.

Theorem 3.3. Let the conditions of Lemma 3.9 be fulfilled and let $x^0\in\partial G$ be a smoothness point of ∂G whereby $u(x^0)\neq 0$. If for a multi-index $\alpha=(\alpha_1,\dots,\alpha_n)\in\mathbb{Z}_+^n$ with $|\alpha|\ge n-\nu$, $1\le|\alpha|\le m$, represented as $\alpha=e^i+\alpha'$, $1\le i\le n$, we have $\omega_i(x^0)\neq 0$ and

$$w_{|\alpha|-(n-\nu)}(x)\left|\int_{\partial G} D_x^{\alpha'}K(x,y)dS_y\right| \ge c_0>0 \quad \text{as} \quad x\to x^0 \tag{3.35}$$

then $u(x^0)w_i(x^0)\int_{\partial G} D_x^{\alpha'}K(x,y)dS_y$ is the main singular part of $D^\alpha u(x)$ as $x\to x^0$ and

$$|D^\alpha u(x)| \ge c_1 \left\{\begin{matrix} 1+|\log\rho(x)|, & |\alpha|=n-\nu \\ \rho(x)^{n-\nu-|\alpha|}, & |\alpha|>n-\nu \end{matrix}\right\} \quad \text{as} \quad x\to x^0$$

where c_1 is a positive constant.

Comparing this with (2.12) we can state that the singularity of $D^\alpha u$ is sharply described. Thus, condition (3.35) or an equivalent condition

$$\Big| \int_{\partial G\cap B(x^0,r^0)} D_x^{\alpha'} K(x,y)\,dS_y \Big| \geq c_0 \begin{Bmatrix} 1+|\log\rho(x)|, & |\alpha|=n-\nu \\ \rho(x)^{n-\nu-|\alpha|}, & |\alpha|>n-\nu \end{Bmatrix} \text{ as } x\to x^0 \quad (3.36)$$

with a fixed $r_0>0$ turns to be deciding for the sharpness of Theorem 3.1.

3.10. Sharpness of Theorem 3.1 (analysis of model examples). Consider the case $n=2$ and, for the simplicity, a region $G\subset\mathbb{R}^2$ containing the interval $(-1,1)$ of the x_1 axis as a part of ∂G (see Figur 3.4). We put $x^0=(0,0)$ and tend

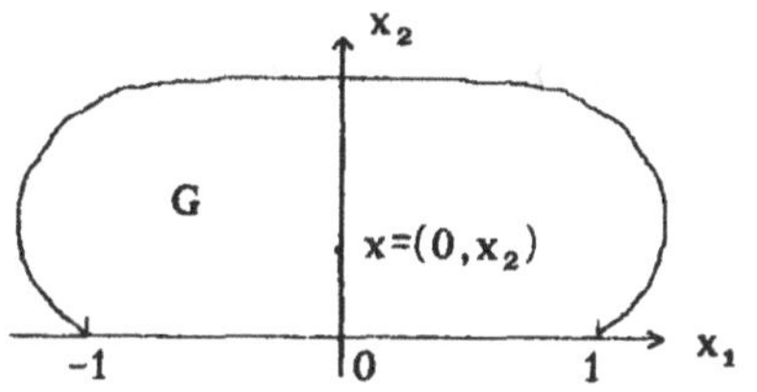

Fig. 3.4

$x=(0,x_2)$ to x^0, i.e. $x_2\to 0$. For derivatives $\partial^k u/\partial x_2^k$, $k\geq\min\{1, 2-\nu\}$, condition (3.36) takes the form

$$\Big|\int_{-1}^{1}\Big[\frac{\partial^{k-1}K(x,y)}{\partial x_2^{k-1}}\Big]_{x_1=0,\,y_2=0} dy_1\Big| \geq c_0 \begin{Bmatrix} 1+|\log x_2|, & k=2-\nu \\ x_2^{2-\nu-k} &, k>2-\nu \end{Bmatrix} \text{ as } x_2\to 0 \quad (3.37)$$

where c_0 is a positive constant. For simplicity, we restrict ourselves to $k=1,2$.

3.10.1. Kernel $K(x,y)=|x-y|^{-\nu}$, $0<\nu<2$. We have

$$\partial K(x,y)/\partial x_2 = -\nu|x-y|^{-\nu-2}(x_2-y_2)$$

and

$$\Big|\int_{-1}^{1}\Big[\frac{\partial K(x,y)}{\partial x_2}\Big]_{x_1=0,\,y_2=0} dy_1\Big| = \nu x_2 \int_{-1}^{1}(x_2^2+y_1^2)^{-(\nu+2)/2}dy_1$$

$$= 2\nu x_2\int_0^1 (x_2^2+y_1^2)^{-(\nu+2)/2}dy_1 \sim x_2\int_0^1(x_2+y_1)^{-\nu-2}dy_1$$

$$= \frac{x_2}{\nu+1}\Big[x_2^{-\nu-1}-(x_2+1)^{-\nu-1}\Big] \sim x_2^{-\nu} \quad \text{as} \quad x_2\to 0$$

where the designation $a\sim b$ as $x_2\to 0$ means that there exist positive constants c_1 and c_2 such that $c_1\leq a(x_2)/b(x_2)\leq c_2$ as $x_2\to 0$. Thus, (3.37) holds for $k=2$, and the estimate

$$|\partial^2 u(x)/\partial x_2^2| \le \text{const}\; x_2^{-\nu} \quad \text{as} \quad x_2 \to 0$$

given by Theorem 3.1 is sharp (assuming that $u(0) \ne 0$). The first derivative $\partial u/\partial x_1$ may have a singularity if $1 \le \nu < 2$. The reader can easily check that (3.37) is fulfilled again:

$$\Big|\int_{-1}^{1}\Big[K(x,y)\Big]_{x_1=0,\,y_2=0} dy_1\Big| \sim \left\{\begin{matrix} 1+|\log x_2|, & \nu=1 \\ x_2^{1-\nu} & , \nu>1 \end{matrix}\right\} .$$

3.10.2. Kernel $K(x,y) = (x_2-y_2)/|x-y|$. It is easy to check that this kernel satisfies (3.2) with $\nu = 0$. The first derivatives of u, the solution to (3.1), are bounded; to examine the sharpness of the estimate

$$|\partial^2 u(x)/\partial x_2^2| \le \text{const}\; (1+|\log x_2|) \quad \text{as} \quad x_2 \to 0$$

given by Theorem 3.1 we check (3.37) with $k=2$, $\nu=0$. We have

$$\partial K(x,y)/\partial x_2 = |x-y|^{-3}(x_1-y_1)^2,$$

$$\Big|\int_{-1}^{1}\Big[\frac{\partial K(x,y)}{\partial x_2}\Big]_{x_1=0,\,y_2=0} dy_1\Big| = \int_{-1}^{1}(x_2^2+y_1^2)^{-3/2}y_1^2\,dy_1 = 2\int_{0}^{1}(x_2^2+y_1^2)^{-3/2}y_1^2\,dy_1 .$$

Representing

$$(x_2^2+y_1^2)^{-3/2}y_1^2 = (x_2^2+y_1^2)^{-1/2} - (x_2^2+y_1^2)^{-3/2}x_2^2$$

we find that

$$\int_0^1 (x_2^2+y_1^2)^{-1/2}dy_1 \sim \int_0^1 (x_2+y_1)^{-1}dy_1 = \log(x_2+1) - \log x_2 \sim |\log x_2|,$$

$$x_2^2\int_0^1 (x_2^2+y_1^2)^{-3/2}dy_1 \sim x_2^2\int_0^1 (x_2+y_1)^{-3}dy_1 = \frac{1}{2}\,x_2^2(x_2^{-2}-(x_2+1)^{-2}) \sim 1 \quad \text{as} \quad x_2 \to 0,$$

and (3.37) with $k=2$, $\nu=0$ is fulfilled.

3.10.3. Kernel $K(x,y) = \log|x-y|$ $(\nu=0)$ represents an example where (3.37) is violated for all $k \ge 2$ and Theorem 3.1 provides non-sharp estimates of the derivatives of a solution to (3.1). Indeed, $\partial K(x,y)/\partial x_2 = |x-y|^{-2}(x_2-y_2)$ and

$$\Big|\int_{-1}^{1}\Big[\frac{\partial K(x,y)}{\partial x_2}\Big]_{x_1=0,\,y_2=0} dy_1\Big| = x_2\int_{-1}^{1}(x_2^2+y_1^2)^{-1}dy_1 \sim 2x_2\int_0^1 (x_2+y_1)^{-2}dy_1$$

$$= 2x_2(x_2^{-1}-(x_2+1)^{-1}) \sim 1 \quad \text{as} \quad x_2 \to 0$$

instead of the logarithmic singularity in (3.37) for $k=2$, $\nu=0$. The reader can continue with $k=3,4,\ldots$. In the next section we consider this example in a more general setting.

3.11. Two dimensional integral equation with a logarithmically singular kernel. We present here a result of U. Kangro (1990a).

Theorem 3.4. Let $G \subset \mathbb{R}^2$ be an open bounded set with piecewise smooth boundary ∂G. Assume that $f \in C^2(\overline{G})$ and

$$K(x,y) = a(x,y) \log|x-y|$$

where $a \in C^2(\overline{G} \times \overline{G})$. If equation (3.1) has a solution $u \in L(G)$ then $u \in C^1(\overline{G}) \cap C^2(G)$ whereby, at any smoothness point of ∂G, the second derivatives of u are continuous up to the boundary, whereas in a neighbourhood of a simple corner point $y^* \in \partial G$,

$$\frac{\partial^2 u(x)}{\partial x_j^2} = (-1)^j u(y^*) a(y^*,y^*) [(\omega_1 \omega_2 - \omega_1' \omega_2') \log|x-y^*|$$

$$+ (\omega_1^2 - \omega_1'^2) \arg_*(x)] + v_j(x), \quad j=1,2,$$

$$\frac{\partial^2 u(x)}{\partial x_1 \partial x_2} = u(y^*) a(y^*,y^*) [(\omega_1^2 - \omega_1'^2) \log|x-y^*| - (\omega_1 \omega_2 - \omega_1' \omega_2') \arg_*(x) + v(x)$$

where v and v_j are continuous functions on the intersection of the neighbourhood of y^* with $\overline{G}$; $\omega = (\omega_1, \omega_2)$ and $\omega' = (\omega_1', \omega_2')$ are the one side limit values of $\omega(y)$, the inner unit normal to ∂G, as $y \to y^*$ so that G is on left, respectively, on the right from $y \in \partial G$;

$$\arg_*(x) = \mathrm{Arg}[x_1 - y_1^* + i(x_2 - y_2^*)], \qquad i = \sqrt{-1},$$

whereby the single-valuedness of this function is achieved cutting the complex plane along a curve which begins at y^* and to the extent of the neighbourhood of y^*, lies in $\mathbb{R}^2 \setminus G$.

Let us make more precise the assumption about the piecewise smoothness of ∂G in Theorem 3.4. It sufficies to assume that ∂G is piecewise Liapunov, i.e. may be represented as a finite union of $C^{1,\mu}$ continuous curves, $\mu > 0$, which may meet each other only at their end points. For a description of the behaviour of the derivatives of the order $k > 2$ the smoothness conditions to the smooth parts of ∂G must be strengthened.

The main qualitative consequence of Theorem 3.4 can be formulated as follows: in the case of the logarithmical kernel, $n = 2$, the (logarithmical) singularities of the second derivatives of a solution to integral equation (3.1) allowed by Theorem 3.1 actually can occur only at the singular points of the boundary. In the second work on this topics, U. Kangro (1992) has shown that this phenominon remains true for the higher order derivatives, too.

The proof of Theorem 3.4 and its extension to higher order derivatives is based on the observation that $\log|x-y^*|$ and $\arg_*(x)$ are conjugate function, i.e. $w = \log|x-y^*| + i\arg_*(x)$ is holomorph. Using the Cauchy–

Riemann formulas, the normal differention of one of those functions can be replaced by tangential differentiation of the other one. We refer to original works of U. Kangro (1990a,1992) for the details.

3.12. Exercises. 3.12.1. Prove that the kernel $K(x,y)=a(x,y)|x-y|^{-\nu}$, $0<\nu<n$, satisfies (3.2) provided that $a(x,y)$ is m times continuously differentiable on $(G\times G)\setminus\{x=y\}$ and

$$|D_x^\alpha D_{x+y}^\beta a(x,y)| \le b'|x-y|^{-|\alpha|} \qquad (|\alpha|+|\beta|\le m),\ b'=\text{const}.$$

3.12.2. Prove that the kernel $K(x,y)=a(x,y)\log|x-y|$ satisfies (3.2) with $\nu=0$ provided that $a(x,y)$ is m times continuously differentiable on $(G\times G)\setminus\{x=y\}$ and

$$|D_x^\alpha D_{x+y}^\beta a(x,y)| \le b'|x-y|^{-\alpha}[1+|\log|x-y||]^{-1} \qquad (|\alpha|+|\beta|\le m).$$

3.12.3. Prove that the kernel $K(x,y)=\varkappa(x,y,|x-y|)$ satisfies (3.2) provided that $\varkappa: G\times G\times\mathbb{R}_+\to\mathbb{R}$ is m times continuously differentiable and

$$\left|D_x^\alpha D_y^\beta \frac{\partial^k}{\partial r^k}\varkappa(x,y,r)\right| \le b'r^{-\nu-k} \qquad (|\alpha|+|\beta|+k\le m)$$

where $0<\nu<n$.

3.12.4. Prove that the Peierls kernel satisfies (3.2) with $\nu=2$ provided that $G\subset\mathbb{R}^3$ is convex and $\sigma,\sigma_0\in BC^m(G)$.

3.12.5. Prove that Theorem 3.1 remains true for the integral equation (3.1) with the kernel

$$K(x,y)=a(x)\mathcal{H}(x,y)b(y)$$

where $\mathcal{H}(x,y)$ satisfies (3.2) and $a,b\in C^{m,\nu}(G)$.

3.12.6. Prove that Theorems 3.1 and 3.2 remain true in the case of an operator valued kernel $K(x,y)\in\mathcal{L}(E,E)$, $x,y\in G$, where E is a Banach space.

4. OUTLINES OF THE DISCRETE CONVERGENCE THEORY

The theory of discrete convergence provides a convenient language for the convergence analysis of discretization methods for integral equations and other problems. In this Chapter we present a self-contained elementary exposition of results concerning the discretization of the linear equations of the type $u = Tu + f$ where T is a compact operator in a Banach space; results concerning the eigenvalue problem for T and non-linear equations are presented without proofs.

For more complete expositions of the theory, with more general problem settings, we refer to Vainikko (1976 a, b, 1978, 1981a).

4.1. Discrete convergence and discrete compactness of a family of elements. Let E and E_h, $0 < h < \bar{h}$, be Banach spaces (all real or all complex) and let $p_h \in \mathcal{L}(E, E_h)$ be so-called connection operators satisfying the condition

$$\|p_h u\|_{E_h} \to \|u\|_E \quad \text{as} \quad h \to 0, \qquad \forall u \in E. \tag{4.1}$$

In applications E_h usually are finite dimensional; in the formal theory this assumption is not necessary.

Definition 4.1. A family $(u_h)_{0<h<\bar{h}}$ of element $u_h \in E_h$ is called discretely converging to an element $u \in E$ if

$$\|u_h - p_h u\|_{E_h} \to 0 \quad \text{as} \quad h \to 0;$$

we write $u_h \dashrightarrow u$. In a similar way the discrete convergence of sequences is defined: a sequence (u_{h_n}) of $u_{h_n} \in E_{h_n}$ with $h_n \to 0$ discretely converges to $u \in E$ if $\|u_{h_n} - p_{h_n} u\|_{E_{h_n}} \to 0$ as $n \to \infty$.

Definition 4.2. A family $(u_h)_{0<h<\bar{h}}$ of elements $u_h \in E_h$ is called discretely compact if any sequence (u_{h_n}), formed by the elements of the family with $h_n \to 0$, contains a discretely convergent subsequence.

Note that the concepts of discrete convergence and discrete compactness depend on the family $\mathcal{P} = (p_h)$ of the connection operators; we assume throughout that $\mathcal{P}$ is fixed.

Using the principle of uniform boundness, it is easy to follow from (4.1) that

$$\limsup_{h\to 0} \|p_h\|_{\mathcal{L}(E,E_h)} < \infty \tag{4.2}$$

(see exercise 4.7.1). The following properties of the discrete convergence and discrete compactness are direct consequences of the definitions, (4.1) and (4.2):

$$u_h \dashrightarrow u,\ u_h \dashrightarrow u' \quad \Rightarrow \quad u = u';$$

$$u_h \dashrightarrow u,\ u_h' \dashrightarrow u',\ \lambda, \lambda' \in \mathbb{R}\ (\text{or } \mathbb{C}) \quad \Rightarrow \quad \lambda u_h + \lambda' u_h' \dashrightarrow \lambda u + \lambda' u';$$

$$u_h \dashrightarrow u \quad \Rightarrow \quad \|u_h\| \to \|u\|;$$

$$u_h \dashrightarrow 0_E \quad \Leftrightarrow \quad \|u_h\| \to 0;$$

$$\forall u \in E,\ p_h u \dashrightarrow u;$$

$$(u^{(h)}) \subset E,\quad \|u^{(h)} - u\| \to 0 \quad \Rightarrow \quad p_h u^{(h)} \dashrightarrow u;$$

$$u_h \dashrightarrow u \quad \Rightarrow \quad (u_h) \text{ discretely compact};$$

$$(u_h) \text{ discretely compact} \quad \Rightarrow \quad \limsup_{h\to 0} \|u_h\| < \infty;$$

$$(u_h), (u_h') \text{ discretely compact},\ \lambda, \lambda' \in \mathbb{R}\ (\text{or } \mathbb{C}) \quad \Rightarrow \quad (\lambda u_h + \lambda' u_h') \text{ discretely compact};$$

$$(u^{(h)}) \subset E \text{ relatively compact} \quad \Rightarrow \quad (p_h u^{(h)}) \text{ discretely compact}. \tag{4.3}$$

4.2. Example. The following example of the discrete convergence will be used in the next chapters. Let $G \subset \mathbb{R}^n$ be an open bounded set and $\Xi_h = \{\xi_{j,h}\}_{j=1}^{l_h}$ a grid for G (a finite number of points of G) such that

$$\operatorname{dist}(x, \Xi_h) \to 0 \quad \text{as} \quad h \to 0, \quad \forall x \in G.$$

Introduce the space $E = BC(G)$ (see Section 2.1) with the norm

$$\|u\|_E = \sup_{x\in G} |u(x)|, \quad u \in E = BC(G),$$

and the space $E_h = C(\Xi_h)$ of functions on the grid Ξ_h with the norm

$$\|u_h\|_{E_h} = \max_{1\le j\le l_h} |u_h(\xi_{j,h})|, \quad u_h \in E_h = C(\Xi_h).$$

Define $p_h \in \mathcal{L}(E, E_h)$ as the operator restricting functions $u \in E$ to the grid Ξ_h: $p_h u \in E_h$ is grid function with the values

$$(p_h u)(\xi_{j,h}) = u(\xi_{j,h}), \quad j = 1, \dots l_h.$$

Then the discrete convergence $u_h \dashrightarrow u$ means that

$$\max_{1\le j\le l_h} |u_h(\xi_{j,h}) - u(\xi_{j,h})| = \|u_h - p_h u\|_{E_h} \to 0 \quad \text{as} \quad h \to 0.$$

Let us check that condition (4.1) is fulfilled in this example. First, $\|p_h u\|_{E_h} \le \|u\|_E$, $u \in E$, and we have (4.2) in a strengthened form:

$$\|p_h\|_{\mathcal{L}(E,E_h)} = 1. \tag{4.4}$$

Further, given a function $u \in BC(G)$ and an $\varepsilon > 0$, take a point $x' \in G$ such that $|u(x')| > \|u\| - \varepsilon$. There is a $\delta > 0$ such that $x \in G$, $|x - x'| < \delta$ imply $|u(x) - u(x')| < \varepsilon$. For sufficiently small h, say for $0 < h < h_\delta$, there is a grid point $\xi_{j,h} \in \Xi_h$ such that $|\xi_{j,h} - x'| < \delta$. Thus,

$$|u(\xi_{j,h})| \ge |u(x')| - |u(\xi_{j,h}) - u(x')| \ge \|u\| - 2\varepsilon$$

and

$$\|p_h u\|_{E_h} \ge \|u\|_E - 2\varepsilon \quad \text{for } 0 < h < h_\delta.$$

Together with (4.4) this provides (4.1).

4.3. Discrete convergence of a family of linear operators. Let us return to the abstract setting of Section 4.1.

<u>Definition</u> 4.3. A family $(T_h)_{0<h<\bar{h}}$ of linear bounded operators $T_h \in \mathcal{L}(E_h, E_h)$ is called discretely converging to $T \in \mathcal{L}(E,E)$ if the following implication holds:

$$E_h \ni u_h \dashrightarrow u \in E \;\Rightarrow\; T_h u_h \dashrightarrow Tu; \tag{4.5}$$

we write $T_h \dashrightarrow T$.

$$\begin{array}{ccc} u \in E & \xrightarrow{\quad T \quad} & E \ni Tu \\ \big\downarrow p_h & & \big\downarrow p_h \\ u_h \in E_h & \xrightarrow{\quad T_h \quad} & E_h \ni T_h u_h \end{array}$$

<u>Lemma</u> 4.1. If $T_h \dashrightarrow T$ then, for any $u \in E$,

$$\|T_h p_h u - p_h T u\| \to 0 \quad \text{as} \quad h \to 0 \tag{4.6}$$

and

$$\limsup_{h \to 0} \|T_h\| < \infty. \tag{4.7}$$

Conversely, if (4.6) holds for any $u \in \mathcal{M} \subseteq E$ where $\mathcal{M}$ is dense in E and (4.7) is true then $T_h \dashrightarrow T$.

<u>Proof</u>. Let $T_h \dashrightarrow T$. Since $p_h u \dashrightarrow u$, (4.5) immediately yields (4.6). Assume that (4.7) is false. Then there are $h_n \to 0$ and $u'_{h_n} \in E_{h_n}$ such that $\|u'_{h_n}\| = 1$, $\|T_{h_n} u'_{h_n}\| \to \infty$ as $n \to \infty$. For $u''_{h_n} = u'_{h_n} / \|T_{h_n} u'_{h_n}\|$ we have $\|u''_{h_n}\| \to 0$, $\|T_{h_n} u''_{h_n}\| = 1$, thus $u''_{h_n} \dashrightarrow 0_E$ but $T_{h_n} u''_{h_n} \dashrightarrow T0_E = 0_F$ does not hold. This contradicts (4.5) and proves (4.7). Note that from (4.5) a similar implication for sequences follows:

$$u_{h_n} \dashrightarrow u \quad (h_n \to 0) \quad \Rightarrow \quad T_{h_n} u_{h_n} \dashrightarrow Tu$$

(one may complete the sequence (u_{h_n}) up to a family (u_h) putting $u_h = p_h u$ for h not involved by (h_n), and then use (4.5)).

Now we prove the inverse statement. Let $u_h \dashrightarrow u$; we have to check that $T_h u_h \dashrightarrow Tu$. Fix an arbitrary $\varepsilon > 0$. Since $\mathcal{U} \subset E$ is dense, there is a $u' \in \mathcal{U}$ such that $\|u' - u\| < \varepsilon$. We have

$$T_h u_h - p_h Tu = T_h(u_h - p_h u) + T_h p_h(u - u') + (T_h p_h u' - p_h Tu') + p_h T(u' - u)$$

and

$$\|T_h u_h - p_h Tu\| \le \|T_h\| \|u_h - p_h u\| + \|T_h p_h u' - p_h Tu'\| + \|p_h\| (\|T_h\| + \|T\|)\varepsilon.$$

Taking into account (4.6), (4.7) and (4.2) we obtain

$$\limsup_{h \to 0} \|T_h u_h - p_h Tu\| \le \text{const} \cdot \varepsilon$$

Since $\varepsilon > 0$ is arbitrary, this means that $\|T_h u_h - p_h Tu\| \to 0$ as $h \to 0$, i.e. $T_h u_h \dashrightarrow Tu$.

<u>Definition</u> 4.4. We say that the discrete convergence $T_h \dashrightarrow T$ is compact, or $T_h \dashrightarrow T$ compactly, if (4.5) and the following implication hold:

$$\limsup_{h \to 0} \|u_h\| < \infty \quad \Rightarrow \quad (T_h u_h) \text{ discretely compact.} \tag{4.8}$$

Let us denote by I and I_h the identity operators in E and E_h, respectively.

<u>Lemma</u> 4.2. Assume that $T_h \dashrightarrow T$ compactly whereby the operators $T_h \in \mathcal{L}(E_h, E_h)$ are compact and the homogeneous equation $v = Tv$ has in E only the trivial solution $v = 0_E$. Then there exists a $h_* > 0$ such that, for $0 < h < h_*$, the operators $I_h - T_h \in \mathcal{L}(E_h, E_h)$ are invertible and

$$\limsup_{h \to 0} \|(I_h - T_h)^{-1}\| < \infty. \tag{4.9}$$

Proof. Suppose that such a h_* does not exist. Then there are $u'_{h_n} \in E_{h_n}$ $(n \in \mathbb{N})$ such that $h_n \to 0$, $\|u'_{h_n}\| = 1$, $\|u'_{h_n} - T_{h_n} u'_{h_n}\| \to 0$ as $n \to \infty$. Define $u'_h = 0$ for $h \ne h_n$ $(n \in \mathbb{N})$. The family (u'_h) is bounded, and due to (4.8), $(T_h u'_h)$ is discretely compact. Hence the sequence $(T_{h_n} u_{h_n})$ contains a discretely converging subsequence: $u'_{h_n} \dashrightarrow u' \in E$ $(n \in \mathbb{N}' \subseteq \mathbb{N})$. Then $\|u'\| = 1$, $u'_{h_n} - T_{h_n} u'_{h_n} \dashrightarrow u' - Tu'$ $(n \in \mathbb{N}')$. On the other hand, $u'_{h_n} - T_{h_n} u'_{h_n} \dashrightarrow 0_E$, therefore $u' - Tu' = 0_E$. This contradicts the condition of the Lemma and proves (4.9).

4.4. Discrete convergence of approximate solutions. Consider the equation

$$u = Tu + f \tag{4.10}$$

where $f \in E$ and $T \in \mathcal{L}(E,E)$. We approximate (4.10) by the equations

$$u_h = T_h u_h + p_h f, \quad 0<h<\bar{h}, \tag{4.11}$$

where $T_h \in \mathcal{L}(E_h, E_h)$. We are interested in the discrete convergence $u_h \dashrightarrow u$ where $u \in E$ and $u_h \in E_h$ are the solutions of equations (4.10) and (4.11), respectively.

Theorem 4.1. Assume that $T_h \dashrightarrow T$ compactly whereby $T_h \in \mathcal{L}(E_h, E_h)$ and $T \in \mathcal{L}(E,E)$ are compact operators. Suppose that the homogeneous equation $v = Tv$ has in E only the trivial solution $v = 0_E$. Then equation (4.10) has a unique solution $u \in E$ and there exists a $h_0 > 0$ such that, for $0<h<h_0$, equation (4.11) has a unique solution $u_h \in E_h$ whereby $u_h \dashrightarrow u$ with the error estimate

$$c_1 e_h \leq \|u_h - p_h u\|_{E_h} \leq c_2 e_h, \quad 0<h<h_0, \tag{4.12}$$

where $e_h = \|T_h p_h u - p_h Tu\|_{E_h}$ and c_1 and c_2 are positive constants not depending on h and f.

Proof. The unique solvability of (4.10) follows from our assumptions concerning T. Further, due to Lemma 4.1, there are $h_1 > 0$ and $c_1 > 0$ such that

$$\|I_h - T_h\| \leq 1/c_1 \quad (0<h<h_1). \tag{4.13}$$

Due to Lemma 4.2, there are $h_0 \in (0, h_1]$ and $c_2 > 0$ such that the operator $I_h - T_h$ is invertible for $0<h<h_0$ and

$$\|(I_h - T_h)^{-1}\| \leq c_2 \quad (0<h<h_0). \tag{4.14}$$

For $0<h<h_0$, equation (4.11) is uniquely solvable. For the solutions of equations (4.10) and (4.11) we have

$$(I_h - T_h)(u_h - p_h u) = T_h p_h u - p_h Tu,$$

$$u_h - p_h u = (I_h - T_h)^{-1}(T_h p_h u - p_h Tu).$$

Using (4.13) and (4.14) we get error estimate (4.12). The discrete convergence $T_h \dashrightarrow T$ implies $\|T_h p_h u - p_h Tu\| \to 0$ (see Lemma 4.1) From (4.12) we obtain that $\|u_h - p_h u\| \to 0$ as $h \to 0$, i.e. $u_h \dashrightarrow u$.

4.5. Convergence and convergence rates in eigenvalue problem. In this Section we assume that Banach spaces E and E_h, $0<h<\bar{h}$, are complex. Let $T \in \mathcal{L}(E,E)$ be compact. Then its spectrum $\sigma(T)$ is made up of 0 (assuming

E to be infinite dimensional) and of at most a countable set of eigenvalues which can accumulate only at 0. For an eigenvalue $\lambda_0 \in \sigma(T)$, $\lambda_0 \neq 0$, let us introduce the eigenspace

$$V(\lambda_0, T) = \{u \in E: \lambda_0 u = Tu\} \subset E$$

and the generalized eigenspace

$$W(\lambda_0, T) = \bigcup_{k \geq 1} \{u \in E: (\lambda_0 I - T)^k u = 0\} \subset E.$$

Due to the compactness of T, these subspaces are finite dimensional. Let us denote by $\varkappa = \varkappa(\lambda_0, T)$ the ascent of λ_0, i.e. the smallest natural number satisfying

$$W(\lambda_0, T) = \{u \in E: (\lambda_0 I - T)^{\varkappa} u = 0\}.$$

Equivalently, $\varkappa$ can be defined as the order of the pole of the resolvent $(\lambda I - T)^{-1}$ at λ_0.

Assuming that $T_h \in \mathcal{L}(E_h, E_h)$ is also compact we introduce, for $0 \neq \lambda_h \in \sigma(T_h)$, the eigenspace

$$V(\lambda_h, T_h) = \{u_h \in E_h: \lambda_h u_h = T_h u_h\} \subset E_h$$

and the generalized eigenspace

$$W(\lambda_h, T_h) = \bigcup_{k \geq 1} \{u_h \in E_h: (\lambda_h I_h - T_h)^k u_h = 0\} \subset E_h.$$

Further, for $0 \neq \lambda_0 \in \sigma(T)$, we denote by $W(\lambda_0, \delta, T_h) \subset E_h$ the linear span of $W(\lambda_h, T_h)$ corresponding to all $\lambda_h \in \sigma(T_h)$ which fall into the circle $C(\lambda_0, \delta) = \{\lambda \in \mathbb{C}: |\lambda - \lambda_0| \leq \delta\}$ where $\delta > 0$ is assumed to be sufficiently small so that $C(\lambda_0, \delta) \cap \sigma(T) = \{\lambda_0\}$. Introduce an average value of $\lambda_h \in \sigma(T_h)$ falling into $C(\lambda_0, \delta)$:

$$\hat{\lambda}_h = \Big(\sum_{\lambda_h \in \sigma(T_h) \cap C(\lambda_0, \delta)} \lambda_h \dim W(\lambda_h, T_h) \Big) \Big/ \dim W(\lambda_0, \delta, T_h).$$

<u>Theorem</u> 4.2. Assume that $T_h \dashrightarrow T$ compactly whereby the operators $T_h \in \mathcal{L}(E_h, E_h)$ and $T \in \mathcal{L}(E, E)$ are compact. Then the following assertions hold.

(i) For any $0 \neq \lambda_0 \in \sigma(T)$ there exists $\lambda_h \in \sigma(T_h)$ such that $\lambda_h \to \lambda_0$ as $h \to 0$. Conversely, if $\sigma(T_h) \ni \lambda_h \to \lambda_0$ as $h \to 0$ then $\lambda_0 \in \sigma(T)$.

(ii) If $\sigma(T_h) \ni \lambda_h \to \lambda_0 \neq 0$ as $h \to 0$ then, for $u_h \in V(\lambda_h, T_h)$, $\|u_h\| = 1$, we have $\operatorname{dist}(u_h, p_h V(\lambda_0, T)) \to 0$ as $h \to 0$, i.e.

$$\inf_{u \in V(\lambda_0, T)} \|u_h - p_h u\| \to 0 \quad \text{as} \quad h \to 0.$$

(iii) For $0 \neq \lambda_0 \in \sigma(T)$ and suffisently small $h > 0$,

$$\dim W(\lambda_o,\delta,T_h) = \dim W(\lambda_o,T)$$

whereby the gap between $W(\lambda_o,\delta,T_h)$ and $p_h W(\lambda_o,T)$ tends to zero, i.e.

$$\Theta(W(\lambda_o,\delta,T_h),p_h W(\lambda_o,T)) := \max\left\{\max_{u_h\in W(\lambda_o,\delta,T_h),\|u_h\|=1} \mathrm{dist}(u_h,p_h W(\lambda_o,T)),\right.$$

$$\left.\max_{u\in W(\lambda_o,T),\|u\|=1} \mathrm{dist}(p_h u, W(\lambda_o,\delta,T_h))\right\} \longrightarrow 0 \quad \text{as} \quad h\to 0.$$

(iv) The convergence rates are estimated as follows: for $\sigma(T_h)\ni\lambda_h \to \lambda_o \in\sigma(T)$, $\lambda_o\neq 0$,

$$\begin{aligned} &|\lambda_h-\lambda_o| \le c e_h^{1/\varkappa}, \\ &\max_{u_h\in V(\lambda_h,T_h),\|u_h\|=1} \mathrm{dist}(u_h,p_h V(\lambda_o,T)) \le c e_h^{1/\varkappa} \end{aligned} \tag{4.15}$$

where c is a constant not depending on h, $\varkappa=\varkappa(\lambda_o,T)$ is the ascent of $\lambda_o\in\sigma(T)$,

$$e_h = \max_{u_o\in W(\lambda_o,T),\|u_o\|=1} \|T_h p_h u_o - p_h T u_o\| \to 0 \quad \text{as} \quad h\to 0.$$

Further, for $0\neq\lambda_o\in\sigma(T)$,

$$\begin{aligned} &|\hat\lambda_h-\lambda_o| \le c e_h, \\ &\Theta(W(\lambda_o,\delta,T_h),p_h W(\lambda_o,T)) \le c e_h. \end{aligned} \tag{4.16}$$

In the case of a simple eigenvalue $\lambda_o\in\sigma(T)$, i.e. when $W(\lambda_o,T)=V(\lambda_o,T)$ is one dimensional, the assertion (iv) reduces to the estimates

$$|\lambda_h-\lambda_o| \le c\|T_h p_h u_o - p_h T u_o\|$$

$$\min_{\mu\in\mathbb{C},|\mu|=1} \|u_h - p_h(\mu u_o)\| \le c\|T_h p_h u_o - p_h T u_o\|$$

where $\lambda_o u_o = T u_o$, $\lambda_h u_h = T_h u_h$, $\|u_o\| = \|u_h\| = 1$.

The assertion (i) and (ii) of Theorem 4.2 are elementary and the reader can easily prove them self. On the other hand, the assertion (iii) and especially (iv) do not belong to elementary ones. We refer to Vainikko (1976 a,b, 1977a) for the proof of Theorem 4.2. There is a possibility to improve the estimates for λ_h and $\hat\lambda_h$ (see Vainikko (1981a), Vainikko and Karma (1988)), e.g. $|\lambda_h-\lambda_o|\le c|\langle T_h p_h u_o - p_h T u_o, u_h^*\rangle|$, $u_h^*\in V(\lambda_h,T_h^*)$, $\|u_h^*\|=1$, in the case of a simple $\lambda_o\in\sigma(T)$. For methods considered in the sequal this improvement does not help.

In Vainikko (1976 a,b, 1977a, 1981 a) the approximation of the eigenvalue problem $Au=\lambda Bu$, $A,B\in\mathcal{L}(E,F)$, is also considered. Karma (1971 a, b, 1983, 1989, 1990) has investigated the approximation of the eigenvalue problem of the type $A(\lambda)u=0$; see Vainikko and Karma (1974 b, 1988), too.

4.6. Approximation of a non-linear equation. Now we consider non-linear equations

$$u = Tu + f \tag{4.17}$$

and

$$u_h = T_h u_h + f_h \tag{4.18}$$

where $f \in E$, $f_h \in E_h$ and $T: \Omega \to E$, $T_h: \Omega_h \to E_h$ are non-linear operators defined on open sets $\Omega \subset E$ and $\Omega_h \subset E_h$, respectively:

$$\begin{array}{ccc} E \supset \Omega & \xrightarrow{\;\;T\;\;} & E \\ \big\downarrow p_h & & \big\downarrow p_h \\ E_h \supset \Omega_h & \xrightarrow{\;\;T_h\;\;} & E_h \end{array}$$

We recall that $T: \Omega \to E$ is called Fréchet differentiable at $u^0 \in \Omega$ if there exists a linear operator $T'(u^0) \in \mathcal{L}(E,E)$ such that

$$\|Tu - Tu^0 - T'(u^0)(u-u^0)\| / \|u-u^0\| \to 0 \quad \text{as} \quad \|u-u^0\| \to 0.$$

<u>Theorem</u> 4.3. Let the following conditions be fulfilled:

(i) equation (4.17) has a solution $u^0 \in \Omega$, and the operator T is Fréchet differentiable at u^0;

(ii) there is a positive δ such that the operator T_h, $0<h<\bar{h}$, is Fréchet differentiable in the ball $\|u_h - p_h u^0\| \le \delta$ of E_h, and for any $\varepsilon>0$ there is a δ_ε, $0<\delta_\varepsilon \le \delta$, such that, for all $h \in (0,\bar{h})$,

$$\|T_h'(u_h) - T_h'(p_h u^0)\| \le \varepsilon \quad \text{whenever} \quad \|u_h - p_h u^0\| \le \delta_\varepsilon ;$$

(iii) $\|T_h p_h u^0 - p_h T u^0\| \to 0$ as $h \to 0$;

(iv) $T_h'(p_h u^0) \dashrightarrow T'(u^0)$ compactly whereby $T_h'(p_h u^0) \in \mathcal{L}(E_h, E_h)$ is compact and the homogeneous equation $v = T'(u^0)v$ has in E only the trivial solution;

(v) $\|f_h - p_h f\| \to 0$ as $h \to 0$.

Then there exist $h_0>0$ and δ_0 $(0<\delta_0 \le \delta)$ such that, for $0<h<h_0$ equation (4.18) has a unique solution $u_h^0 \in \Omega_h$ in the ball $\|u_h - p_h u^0\| \le \delta_0$. Thereby $u_h^0 \dashrightarrow u^0$ and the error estimate

$$c_1 e_h \le \|u_h^0 - p_h u^0\|_{E_h} \le c_2 e_h, \quad 0<h<h_0, \tag{4.19}$$

holds where $e_h = \|p_h u^0 - T_h p_h u^0 - f_h\|_{E_h} = \|(p_h T u^0 - T_h p_h u^0) + (p_h f - f)\|_{E_h}$ and c_1 and c_2 are positive constant not depending on h.

We refer to Vainikko (1976a, b or 1978) where this Theorem is elementarily proved in a more general setting. One can find there also a convergence theorem concerning a case with non-differentiable operators.

4.7. Exercises. 4.7.1. Prove that (4.1) implies (4.2). Caution: the principle of the uniform boundedness of linear operators in its standard formulation (see e.g. Yosida (1965)) cannot be directly applied since the ranges of $p_h \in \mathcal{L}(E,E_h)$ are in different spaces. Hint: introduce the space $m = m((E_h)_{0<h<\bar{h}})$ of families $\bar{u}=(u_h)$, $u_h \in E_h$, with the norm

$$\|\bar{u}\| = \sup_{0<h<\bar{h}} \|u_h\|_{E_h},$$

and assign to $p_h \in \mathcal{L}(E,E_h)$ the operator $\bar{p}_h \in \mathcal{L}(E,m)$ via the formula

$$\bar{p}_h u = \bar{u} = (u_{h'}) \quad \text{with} \quad u_{h'} = \begin{Bmatrix} p_h u & \text{if } h'=h \\ 0 & \text{if } h' \neq h \end{Bmatrix}, \quad u \in E.$$

4.7.2. Prove the properties of the discrete convergence and of the discrete compactness listed in the end of Section 4.1.

4.7.3. Analyze the concepts of discrete convergence $u \dashrightarrow u$, discrete compactness of (u_h), discrete convergence $T_h \dashrightarrow T$ and compact convergence $T_h \dashrightarrow T$ in the following cases: (i) $E_h = E$, $p_h = I$; (ii) $E_h \subset E$ are subspaces, $p_h \in \mathcal{L}(E,E_h)$ are projectors. In case (i), the concept of the discrete compact convergence $T_h \dashrightarrow T$ is equivalent to the concept of collectively compact approximation introduced by Anselone (1971); see Sobolev (1956), too.

4.7.4. Let $T_h \dashrightarrow T$ compactly and $S_h \dashrightarrow S$ where $T_h, S_h \in \mathcal{L}(E_h,E_h)$, $T,S \in \mathcal{L}(E,E)$. Prove that $S_h T_h \dashrightarrow ST$ compactly, $T_h S_h \dashrightarrow TS$ compactly.

4.7.5. Let $T_h \dashrightarrow T$ compactly and $S_h \dashrightarrow 0$ (null operator). Prove that $\|S_h T_h\| \to 0$.

4.7.6. Assume that $T_h \dashrightarrow T$ compactly and, for $T_h, T_h' \in \mathcal{L}(E_h,E_h)$, we have $\|T_h' - T_h\| \to 0$ as $h \to 0$. Prove that $T_h' \dashrightarrow T$ compactly. Reformulate Theorem 4.1 for the case where the approximating equation is given by $u_h = T_h' u_h + f_h$, $\|f_h - p_h f\| \to 0$ as $h \to 0$.

4.7.7. Prove the assertions (i) and (ii) of Theorem 4.2.

4.7.8. Let the conditions of Theorem 4.3 be fulfilled and let the following hold: T and T_h are defined on the closures $\bar{\Omega}$ and $\bar{\Omega}_h$ respectively; the solution u^0 of equation (4.17) is unique in $\bar{\Omega}$;

$u_h \in \bar{\Omega}_h \ (0<h<\bar{h}) \Rightarrow (T_h u_h)$ is discretely compact;

$h_n \to 0,\ \bar{\Omega}_{h_n} \ni u_{h_n} \dashrightarrow u \Rightarrow u \in \bar{\Omega},\quad T_{h_n} u_{h_n} \dashrightarrow Tu.$

Prove that for sufficiently small $h>0$, the solution u_h^0 of equation (4.18) is unique in $\bar{\Omega}_h$.

5. PIECEWISE CONSTANT COLLOCATION AND RELATED METHODS

This Chapter is devoted to the convergence analysis of the piecewise constant collocation method (PCCM) and related methods for weakly singular integral equations. The methods are constructed so that, in the case of not too strong singularities, they are of the accuracy $\mathcal{O}(h^2)$ where h is the maximal diameter of the cells.

5.1 Assumptions about the boundary. In that follows we assume that $G\subset\mathbb{R}^n$ is an open bounded set with piecewise smooth boundary. More precisely, the following condition is sufficient.

Condition (PS). The boundary ∂G of $G\subset\mathbb{R}^n$ can be covered by a finite number of compact hypersurfaces Γ_i which may meet each other on manifolds of dimensions not exceeding n-2; there exist constants $c_0>0$ and $r_0>0$ such that every piece $\Gamma_i\cap\bar{B}(x,r_0)$, $x\in\Gamma_i$, in a suitable orthogonal system of coordinates obtained from the original system by the translation of the origin to x and a rotation of the axes, is representable in the form

$$z_n=\varphi(z'),\quad z'=(z_1,\dots,z_{n-1})\in Z_{i,x},$$

where $Z_{i,x}\subset\mathbb{R}^{n-1}$ is a closed region and function $\varphi=\varphi_{i,x}$ is continuous on $Z_{i,x}$ and continuously differentiable with $|\text{grad}\ \varphi(z')|\le c_0$ everywhere in $Z_{i,x}$ except a possible manifold of dimension not exceeding n-2.

A ball, cube, cylinder, cone (see Figure 5.1) and sets diffeomorphic to

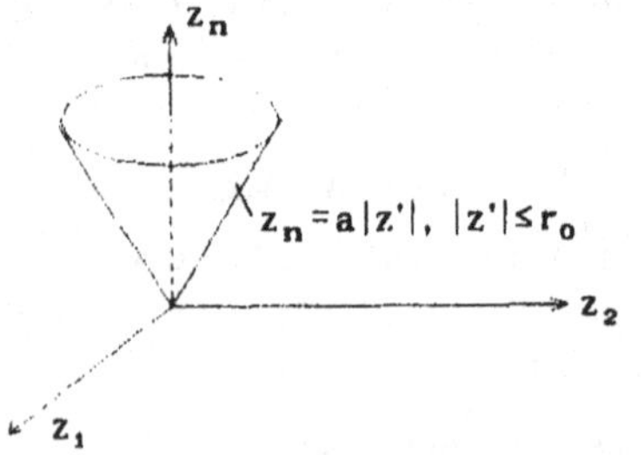

Fig. 5.1

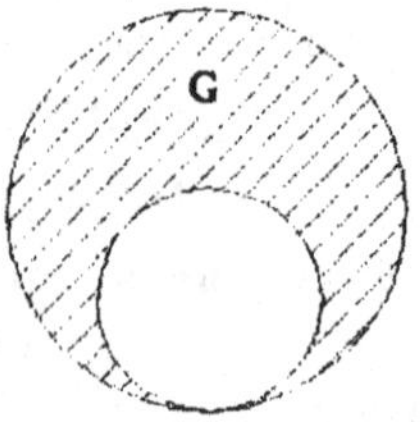

Fig. 5.2.

those have (PS)-boundaries. The region between two tangential balls (see Figure 5.2) has also (PS)-boundary; in this example G does not satisfy the cone condition.

<u>Lemma</u> 5.1. Let ∂G satisfy (PS); denote by

$$\Gamma^h = \{x \in G: \rho(x) < h\}, \; h > 0,$$

the boundary layer of thickness h. Assume that the kernel K(x,y) satisfies condition (3.2) for $|\alpha| = |\beta| = 0$:

$$|K(x,y)| \le b \begin{Bmatrix} 1 & , & \nu < 0 \\ 1 + |\log|x-y||, & & \nu = 0 \\ |x-y|^{-\nu} & , & 0 < \nu < n \end{Bmatrix}.$$

Then

$$\sup_{x \in G} \int_{\Gamma^h} |K(x,y)| dy \le c \varepsilon_{\nu,h}, \qquad \varepsilon_{\nu,h} = \begin{Bmatrix} h & , & \nu < n-1 \\ h(1+|\log h|), & & \nu = n-1 \\ h^{n-\nu} & , & \nu > n-1 \end{Bmatrix}, \quad (5.1)$$

where the constant c is independent of h. Further,

$$\sup_{x \in G} \int_{\Gamma^h \backslash B(x,ah)} |x-y|^{-n} dy \le c_a = \text{const} \quad (a > 0). \quad (5.2)$$

<u>Proof</u>. It sufficies to establish (5.1) for $K(x,y) = |x-y|^{-\nu}$, $0<\nu<n$, since a milder singularity of the kernel can be estimated by $|x-y|^{-\nu}$, $0<\nu<n$, and $\varepsilon_{\nu,h} = h$ for $\nu < n-1$. Since

$$\int_{B(x,ah)} |x-y|^{-\nu} dy = \sigma_n (ah)^{n-\nu}/(n-\nu),$$

we see that (5.1) is equivalent to

$$\sup_{x \in G} \int_{\Gamma^h \backslash B(x,ah)} |x-y|^{-\nu} dy \le c \varepsilon_{\nu,h}, \quad 0 < \nu < n.$$

This inequality and inequality (5.2) will be satisfied if, for any Γ_i introduced in condition (PS), we have

$$\sup_{x \in G} \int_{\Gamma_i^h \backslash B(x,ah)} |x-y|^{-\nu} dy \le c \varepsilon_{\nu,h}, \quad 0 < \nu \le n, \quad (5.3)$$

where

$$\Gamma_i^h = \{x \in G: \text{dist}\,(x, \Gamma_i) < h\}.$$

Let us assign to any $x \in G$ a nearest point x' on Γ_i. Then, for $y \in \Gamma_i^h \backslash B(x,ah)$, we have $|x-y| \ge ah$,

$$|x-x'| \le |x-y| + \text{dist}(y,\Gamma_i) \le |x-y| + h \le (1+a^{-1})|x-y|,$$

$$|x'-y| \le |x'-x| + |x-y| \le (2+a^{-1})|x-y|,$$

$$|x-y|^{-\nu} \le (2+a^{-1})^{\nu}|x'-y|^{-\nu}.$$

Devide the domain of integration in (5.3) into two subdomains

$$D_{1,h} = \Gamma_1^h \setminus [B(x,ah) \cup B(x',ah)], \quad D_{2,h} = [\Gamma_1^h \cap B(x',ah)] \setminus B(x,ah).$$

We have

$$\int_{D_{1,h}} |x-y|^{-\nu} dy \le (2+a^{-1})^{\nu} \int_{\Gamma_1^h \setminus B(x',ah)} |x'-y|^{-\nu} dy,$$

$$\int_{D_{2,h}} |x-y|^{-\nu} dy \le (ah)^{-\nu} \int_{B(x',ah)} dy = \text{const}\, h^{n-\nu} \le \text{const}\, \varepsilon_{\nu,h}.$$

This allows to substitude the supremum over $x \in G$ in (5.3) by the supremum over $x \in \Gamma_1$. To prove the Lemma, we have to show that

$$\sup_{x \in \Gamma_1} \int_{\Gamma_{1,x}^h} |x-y|^{-\nu} dy \le c\, \varepsilon_{\nu,h} \tag{5.3'}$$

where

$$\Gamma_{1,x}^h = \{y \in G: \text{dist}(y,\Gamma_1) < h, \ \ 2h < |x-y| < r_0\}$$

with an arbitrarily fixed r_0; we take as r_0 the number introduced in condition (PS). We fixed $a=2$; note that this does not restrict the generality of assertion (5.2) since

$$\int_{B(x,a_2h) \setminus B(x,a_1h)} |x-y|^{-n} dy = \text{const} \log(a_2/a_1), \quad 0<a_1<a_2.$$

For $x \in \Gamma_1$, $y \in \Gamma_{1,x}^h$, denoting by $y' \in \Gamma_1$ the nearest point to y on Γ_1, we have

$$|x-y'| \le |x-y| + |y-y'| \le |x-y| + h \le 2|x-y|.$$

Therefore, integrating over the direction orthogonal to Γ_1, we obtain (see Figure 5.3)

$$\int_{\Gamma_{1,x}^h} |x-y|^{-\nu} dy \le 2^{\nu} \int_{\Gamma_{1,x}^h} |x-y'|^{-\nu} dy \le c'h \int_{\Gamma_1 \cap [B(x,r_0) \setminus B(x,h)]} |x-y'|^{-\nu} dS_{y'}.$$

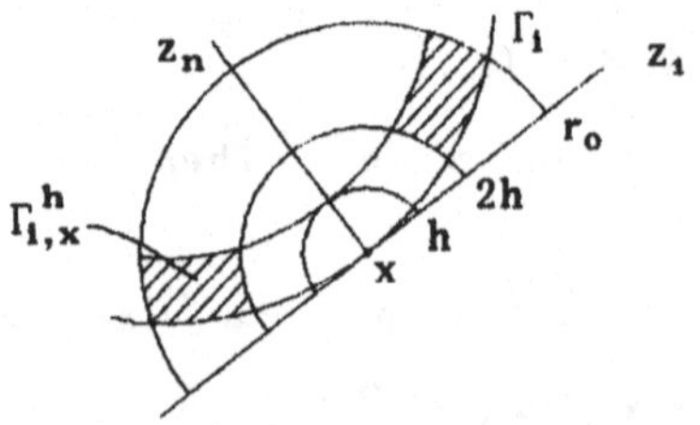

Fig. 5.3

Representing $\Gamma_i \cap B(x,r_0)$ in the system of coordinates exposed in condition (PS), we have

$$\Gamma_i \cap [B(x,r_0)\setminus B(x,h)] = \{z\in\mathbb{R}^n:\ z_n=\varphi(z'),\ z'\in Z_{i,x}\subset\mathbb{R}^{n-1},\ h^2\le|z'|^2+|\varphi(z')|^2\le r_0^2\}$$

$$\subset \{z\in\mathbb{R}^n:\ z_n=\varphi(z'),\ z'\in Z_{i,x}\subset\mathbb{R}^{n-1},\ (1+c_0^2)^{-1/2}h<|z'|<r_0\},$$

$$|x-y'| = (|z'|^2+|\varphi(z')|^2)^{1/2}\ge|z'|,\quad y'\in\Gamma_i\cap B(x,r_0),$$

$$dS_{y'} = (1+|\operatorname{grad}\varphi(z')|^2)^{1/2}dz'\le(1+c_0^2)^{1/2}dz'$$

and

$$\sup_{x\in\Gamma_i}\int_{\Gamma^h_{i,x}}|x-y|^{-\nu}dy\le c'h(1+c_0^2)^{1/2}\int_{\{z'\in\mathbb{R}^{n-1}:\ (1+c_0^2)^{-1/2}h<|z'|<r_0\}}|z'|^{-\nu}dz'$$

$$\le ch\begin{Bmatrix}1, & \nu<n-1\\ 1+|\log h|, & \nu=n-1\\ h^{n-1-\nu}, & \nu>n-1\end{Bmatrix}=c\,\varepsilon_{\nu,h}\,.$$

Thus, (5.3') holds and Lemma 5.1 is proved.

5.2. Partition of G. (i) There are practical and principled reasons to use "approximate" partitions of G (see exercise 2.7.1). For any h ($0<h<\tilde h$), we introduce measurable sets (cells) $G_{j,h}\subset\mathbb{R}^n$ ($j=1,\dots,I_h$) of

$$\operatorname{diam} G_{j,h}\le h \tag{5.4}$$

such that $\operatorname{meas} G^0_{j,h}=\operatorname{meas}\overline{G}_{j,h}$ (i.e. the measures of the interior and the closure of a cell are equal), $G^0_{i,h}\cap G^0_{j,h}=\emptyset$ for $i\ne j$ and

$$(\overline{G}\setminus\overline{G}_h)\cup(\overline{G}_h\setminus\overline{G})\subset\{x\in\mathbb{R}^n:\ \rho(x)\equiv\operatorname{dist}(x,\partial G)\le h^2\}=:S_h \tag{5.5}$$

where

$$G_h=\bigcup_{j=1}^{I_h}G_{j,h}\,.$$

In addition, we assume that in any boundary-incident cell $G_{j,h}$ (i.e. in cells with non-void $\partial G\cap\operatorname{co}G_{j,h}$) one can take a non-void measurable part $G'_{j,h}\subset G_{j,h}\cap G$ such that

$$d_G\text{-}\operatorname{diam}G'_{j,h}\le h \tag{5.6}$$

and

$$G_{j,h}\setminus G'_{j,h}\subset S_h. \tag{5.7}$$

Note that, for inner cells (i.e. for cells with void $\partial G\cap\operatorname{co}G_{j,h}$) one has

$$d_G\text{-diam}\, G_{j,h} = \text{diam}\, G_{j,h} \le h.$$

Here coA stands for the convex span of $A \subset \mathbb{R}^n$ and d_G-diam A stands for the diameter of A with respect to the metric d_G (see Section 2.1). Condition (5.5) means that the approximate partition is of h^2-accuraccy.

In any cell $G_{j,h}$ we choose a collocation point $\xi_{j,h}$:

$$\left.\begin{aligned} &\xi_{j,h} = (\text{meas}\, G_{j,h})^{-1} \int_{G_{j,h}} y\,dy \quad (\text{centroid of } G_{jh}) \text{ if } \text{co}\, G_{j,h} \subset G, \\ &\xi_{j,h} \in G'_{j,h} \quad \text{arbitrary if } \partial G \cap \text{co}\, G_{j,h} \ne \emptyset \end{aligned}\right\} \tag{5.8}$$

(requiring that the centroid of any inner cell belongs to this cell).

(ii) Let us construct an example of partition. Denote by

$$B_{\lambda,h} = \{x \in \mathbb{R}^n:\ \lambda_k h_k < x_k < (\lambda_k+1)h_k,\ \ k=1,\dots,n\}$$

the "boxes" in $\mathbb{R}^n$ of diameter $h=(h_1^2+\dots+h_n^2)^{1/2}$ where h_k is the step size in the direction of kth axis and

$$\lambda = (\lambda_1,\dots,\lambda_n) \in \mathbb{Z}^n$$

is a multiindex with integer components. Denote by Λ_h the set of $\lambda \in \mathbb{Z}^n$ such that $B_{\lambda,h} \cap G$ is non-void. Define

$$G_{\lambda,h} = B_{\lambda,h} \cap G, \quad \lambda \in \Lambda_h. \tag{5.9}$$

For inner cells $G_{\lambda,h}$, their centroids (used as collocation points) are given by

$$\xi_{\lambda,h} = ((\lambda_1+\tfrac{1}{2})h_1,\dots,(\lambda_n+\tfrac{1}{2})h_n); \tag{5.10}$$

for boundary-incident cells, choose $\xi_{\lambda,h} \in G_{\lambda,h}$ arbitrarily. Some trouble can be caused by condition (5.6) if G has an inner boundary - a box incident to the inner boundary ought be devided into two cells, with an arbitrary choice of collocation points in both of them. This is an example of a sharp partition. It remains to renumerate the cells $G_{\lambda,h}$ using one dimensional indexing with $j \in \mathbb{N}$.

If inside of a bondary-incident cell, ∂G is C^2-smooth then the boundary may be rectified using a tangent or secant plane. We obtain an approximate partition of G. There are other possibilities to make the partition more convenient for the practice — omit too small cells (of diameter ch^2), join smaller cells to larger ones e.t.c. (see Figures 5.4, 5.5). In inequalities (5.4) and (5.6) and in the inequality inside (5.5), a constant not depending on h and not influencing on the convergence analysis may appear.

(iii) For $G \subset \mathbb{R}^n$ of not too complicated structure, it is possible to construct partions of G so that only the points of type (5.10) will be used as collocation points. We first fix all points (5.10) falling into G, then take the cells $B_{\lambda,h} \cap G$ around them and at last join the remaining boundary

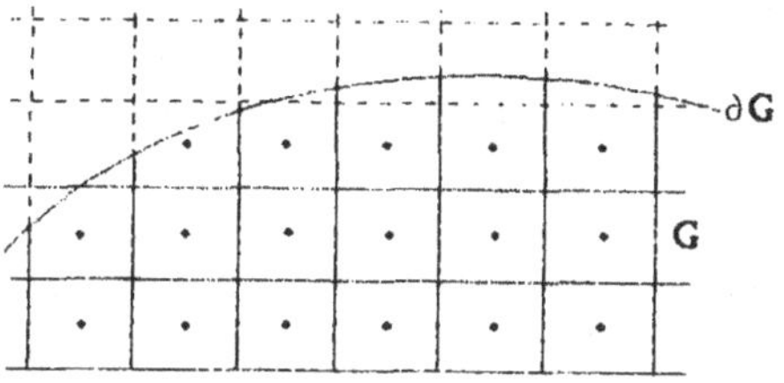

Fig. 5.4

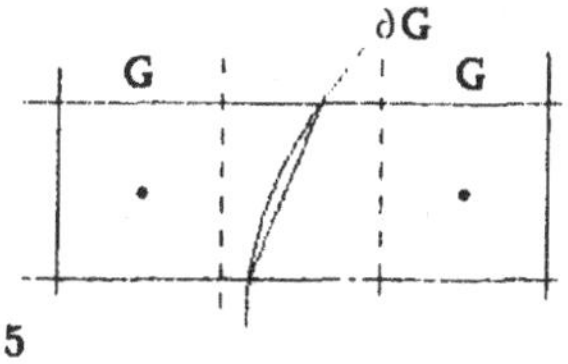

Fig. 5.5

incident cells (with centers of corresponding $B_{\lambda,h}$ outside of G) to their neighbours (see Figure 5.4). Again, some modifications in the construction are possible (see Figure 5.5).

5.3. PCCM and related methods. Starting from the approximate partition of G described in Section 5.2 (i) we represent an approximate solution to integral equation (3.1) as a piecewise constant function

$$\bar{u}_h = \sum_{j=1}^{l_h} u_{j,h} \chi_{j,h}, \qquad \chi_{j,h}(x) = \left\{ \begin{array}{ll} 1, & x \in G_{j,h} \\ 0, & x \notin G_{j,h} \end{array} \right\}.$$

In the integral equation we approximate the domain of integration G by G_h:

$$u(x) = \int_{G_h} K(x,y) u(y) dy + f(x), \qquad G_h = \bigcup_{j=1}^{l_h} G_{j,h}.$$

Replacing u by $\bar{u}_h$ and collocating at points $\xi_{i,h}$ (see (5.8)) we obtain a system of linear algebraic equations to find the values $u_{j,h}$ $(j=1, \dots, l_h)$, namely

$$u_{i,h} = \sum_{j=1}^{l_h} \int_{G_{j,h}} K(\xi_{i,h}, y) dy\, u_{j,h} + f(\xi_{i,h}), \quad i = 1, \dots l_h. \tag{5.11}$$

In general, G_h need not be contained in G, and then we assume that the kernel K(x,y) is extended to $G \times G_h$ so that condition (3.2) remains for $|\alpha| = |\beta| = 0$ valid:

$$|K(x,y)| \le b \left\{ \begin{array}{ll} 1 & , \ \nu<0 \\ 1+|\log|x-y||, & \nu=0 \\ |x-y|^{-\nu} & , \ \nu>0 \end{array} \right\}, \quad x \in G,\ y \in G_h.$$

For practical purposes, sometimes (e.g. in the situation on Figure 5.5) it is expedient to redefenite $K(x,y)$ on some parts of S_h, the h^2 boundary layer around ∂G. This is legitimate if the last inequality remains valid.

We refer to (5.11) as to the piecewise constant collocation method (PCCM). We are mainly interested only in the grid values $u_{j,h} \approx u(\xi_{j,h})$, $j=1,\dots,l_h$; the piecewise constant function $\bar{u}_h$ itself has too bad approximation properties and there is a possibility to define a more accurate approximation on whole G knowing the grid values.

Mostly the integrals in (5.11) cannot be evaluated exactly. Using some standard approximate evaluation procedures we obtain different modifications of the PCCM. Usually the kernel is of the form $K(x,y)=\mathcal{k}(x,y)\varkappa(x,y)$ where $\mathcal{k}(x,y)$ is smooth and the singularity is concentrated in the factor $\varkappa(x,y)$, e.g. $\varkappa(x,y)=\log|x-y|$ or $\varkappa(x,y)=|x-y|^{-\nu}$, $0<\nu<n$. The following modification of PCCM (5.11) is natural:

$$u_{i,h} = \sum_{j=1}^{l_h} \mathcal{k}(\xi_{i,h},\xi_{j,h}) \int_{G_{j,h}} \varkappa(\xi_{i,h},y)\,dy\, u_{j,h} + f(\xi_{i,h}), \quad i=1,\dots l_h. \tag{5.12}$$

In some case, e.g. if $G_{j,h}\subset\mathbb{R}^n$ are rectangles ($n=2$) and $\varkappa(x,y)=\log|x-y|$ or $\varkappa(x,y)=|x-y|^{-1}$, the integrals in (5.12) can be exactly evaluated.

A more standard modification is based on the approximation

$$\int_{G_{j,h}} K(\xi_{i,h},y)\,dy \approx K(\xi_{i,h},\xi_{j,h})\,\text{meas}\,G_{j,h}, \quad i\neq j.$$

Omitting the terms where $\xi_{i,h}$ and $\xi_{j,h}$ are too close to one another we obtain the system

$$u_{i,h} = \sum_{\substack{j=1 \\ \text{dist}(\xi_{i,h},\text{co}G_{j,h})\geq h}}^{l_h} K(\xi_{i,h},\xi_{j,h})\,\text{meas}\,G_{j,h}\,u_{j,h} + f(\xi_{i,h}), \quad i=1,\dots,l_h. \tag{5.13}$$

One can view (5.13) also as a cubature formula method (CFM) to solve (3.1). Namely, approximating the integral in (3.1) by the cubature formula

$$\int_G u(y)\,dy \approx \sum_{j=1}^{l_h} w_{j,h}\,u(\xi_{j,h}), \quad w_{j,h}=\text{meas}\,G_{j,h},$$

collocating at the points $\xi_{i,h}$, $i=1,\dots,l_h$, and rejecting the terms with close arguments, we obtain the same system (5.13).

It is clear that (5.13) is a rather coarse approximation to (5.11). Nevertheless, in many cases, as it will appear in the next Section, both methods are of the same accuracy. A more fine cubature evaluation of integrals will be examined in Chapter 6. Here we present a modification of CFM proposed by L. Kantorovich (see e.g. Kantorovich and Krylov (1962)):

$$u_{i,h} = \sum_{\substack{j=1 \\ j \neq i}}^{l_h} K(\xi_{i,h}, \xi_{j,h}) w_{j,h} u_{j,h} + \beta_{i,h} u_{i,h} + f(\xi_{i,h}), \quad i = 1, \dots, l_h, \tag{5.14}$$

where

$$w_{j,h} = \operatorname{meas} G_{j,h}, \qquad \beta_{i,h} = \int_{G_h} K(\xi_{i,h}, y)\,dy - \sum_{\substack{j=1 \\ j \neq i}}^{l_h} K(\xi_{i,h}, \xi_{j,h}) w_{j,h}.$$

The idea is to mitigate the singularity of the kernel by factor $u(y)-u(x)$ rewriting the integral equation in the form

$$u(x) = \int_{G_h} K(x,y)[u(y)-u(x)]\,dy + u(x) \int_{G_h} K(x,y)\,dy + f(x).$$

Approximating the first integral by the cubature and collocating at $\xi_{i,h}$ we obtain the system

$$u_{i,h} = \sum_{\substack{j=1 \\ j \neq i}}^{l_h} K(\xi_{i,h}, \xi_{j,h}) w_{j,h} (u_{j,h} - u_{i,h}) + u_{i,h} \int_{G_h} K(\xi_{i,h}, y)\,dy + f(\xi_{i,h}),$$

$$i = 1, \dots, l_h,$$

which is equivalent to (5.14).

Note that method (5.14) is of the same complexity as the basic method (5.11) — one needs to evaluate the integrals

$$\int_{G_h} K(\xi_{i,h}, y)\,dy = \sum_{j=1}^{l_h} \int_{G_{j,h}} K(\xi_{i,h}, y)\,dy, \quad i = 1, \dots, l_h.$$

An adventage of method (5.14) appears when the fast Fourier techniques will be involved to solve the systems (see Section 5.14)

5.4. Convergence rates of the methods. At first look it can seem that the methods described in Section 5.3 are at most of $\mathcal{O}(h)$ accuracy since they are based on the piecewise constant approximation of the solution. Actually, due to the special choice of the collocation points, these methods are of $\mathcal{O}(h^2)$ accuracy at those points if the kernel is of not too strong singularity; the strength of the singularity may be different for different methods.

We shall assume that the kernel $K(x,y)$ satisfies (3.2) with $m=2$ and that

$$f \in C^{2,\mu}(G), \quad \nu \leq \mu < n, \tag{5.15}$$

where $\nu \in \mathbb{R}$ $(\nu < n)$ is the number from condition (3.2). Further, we assume that, for any $x^1, x^2 \in G$,

$$|f(x^1)-f(x^2)| \le \text{const}\cdot \left\{ \begin{array}{ll} d_G(x^1,x^2) & ,\ \mu<n-1 \\ d_G(x^1,x^2)[1+|\log d_G(x^1,x^2)|], & \mu=n-1 \\ [d_G(x^1,x^2)]^{n-\mu} & ,\ \mu>n-1 \end{array} \right\}. \tag{5.16}$$

Recall (see Section 2.5) that in many cases (5.16) is a consequence of (5.15).

Now we formulate the main results of the Chapter.

Theorem 5.1. Let the following conditions be fulfilled:

1. $G\subset\mathbb{R}^n$ is open and bounded, ∂G satisfies condition (PS);
2. partitions of G and collocation points satisfy conditions (5.4)-(5.8);
3. kernel $K(x,y)$ satisfies condition (3.2) with $m=2$;
4. f satisfies conditions (5.15) and (5.16);
5. integral equation (3.1) has a unique solution $u\in BC(G)$.

Then there is a $h_0>0$ such that, for $0<h<h_0$, systems (5.11)-(5.14) are uniquely solvable, and the following error estimates hold.

(i) For PCCM (5.11), the estimate

$$\max_{1\le i\le l_h} |u_{i,h}-u(\xi_{i,h})| \le \text{const}\, \varepsilon_{\nu,h}\,\varepsilon_{\mu,h} \tag{5.17}$$

is true where

$$\varepsilon_{\nu,h} = \left\{ \begin{array}{ll} h & ,\ \nu<n-1 \\ h(1+|\log h|), & \nu=n-1 \\ h^{n-\nu} & ,\ \nu>n-1 \end{array} \right\}. \tag{5.18}$$

Thereby, for the function

$$u_h(x) = \sum_{j=1}^{l_h} \int_{G_{j,h}} K(x,y)\,dy\, u_{j,h} + f(x), \quad x\in G, \tag{5.19}$$

we have

$$\sup_{x\in G} |u_h(x)-u(x)| \le \text{const}\, \varepsilon_{\nu,h}\,\varepsilon_{\mu,h}. \tag{5.20}$$

(ii) For method (5.12), estimate (5.17) holds provided that $\mathscr{k}\in BC^2(G\times G)$ and $\varkappa$ satisfies (3.2) with $m=2$. Thereby, for

$$u_h(x) = \sum_{j=1}^{l_h} \mathscr{k}(x,\xi_{j,h}) \int_{G_{j,h}} \varkappa(x,y)\,dy\, u_{j,h} + f(x), \quad x\in G, \tag{5.21}$$

estimate (5.20) holds.

(iii) For CFM (5.13), the estimate

$$\max_{1\le i\le l_h} |u_{i,h}-u(\xi_{i,h})| \le \text{const}\, (\varepsilon_{\nu,h}\,\varepsilon_{\mu,h}+\varepsilon'_{\nu,h}) \tag{5.22}$$

is true where

$$\varepsilon'_{\nu,h} = \left\{ \begin{array}{ll} h^2 & , \quad \nu<n-2 \\ h^2(1+|\log h|), & \nu=n-2 \\ h^{n-\nu} & , \quad \nu>n-2 \end{array} \right\} . \tag{5.23}$$

Thereby, for the function

$$u_h(x) = \sum_{\substack{j=1 \\ \mathrm{dist}(x,\mathrm{co}G_{j,h})\geq h}}^{l_h} K(x,\xi_{j,h})\,\mathrm{meas}\,G_{j,h}\,u_{j,h} + f(x), \quad x\in G, \tag{5.24}$$

we have

$$\sup_{x\in G} |u_h(x)-u(x)| \leq \mathrm{const}\,(\varepsilon_{\nu,h}\varepsilon_{\mu,h}+\varepsilon'_{\nu,h}). \tag{5.25}$$

(iv) For Kantorovich-Krylov modification (5.14), estimate (5.17) holds provided that

$$|\xi_{i,h}-\xi_{j,h}| \geq c_0 h \quad (i\neq j), \quad c_0>0, \tag{5.26}$$

$$d_G(x^1,x^2) \leq \mathrm{const}\,|x^1-x^2|, \quad \forall x^1,x^2\in G \tag{5.27}$$

where the constants are independent of h. Thereby, for function (5.19) with $u_{j,h}$ found from (5.14), estimation (5.20) is true.

Note that condition (5.27) bans the inner boundary.

We see that all four methods are of $\mathcal{O}(h^2)$ accuracy if $\nu<n-2$. For $\nu<n-1$ methods (5.11), (5.12) and (5.14) remain to be of $\mathcal{O}(h^2)$ accuracy but CFM (5.13) is more coarse.

5.5. Block scheme of the proof of Theorem 5.1. We intend to make use of Theorem 4.1. To this end, introduce the spaces $E=BC(G)$, $E_h=C(\Xi_h)$ and connection operators $p_h\in\mathcal{L}(E,E_h)$ as in Section 4.2. Here $\Xi_h=\{\xi_{j,h}\}_{j=1}^{l_h}$ is the set of collocation points (5.8). We treat (3.1) as equation

$$u = Tu+f$$

in the space $E=BC(G)$ defining $T\in\mathcal{L}(E,E)$ via the formula

$$(Tu)(x) = \int_G K(x,y)\,u(y)\,dy\,, \quad u\in E,\ x\in G. \tag{5.28}$$

Systems (5.11)-(5.14) can be treated, respectively, as equations

$$u_h = T_h u_h + p_h f, \qquad u_h = T'_h u_h + p_h f,$$

$$u_h = T''_h u_h + p_h f, \qquad u_h = T'''_h u_h + p_h f$$

in the space $E_h=C(\Xi_h)$ where T_h, T'_h, T''_h, $T'''_h\in\mathcal{L}(E_h,E_h)$ are the corresponding discretizations of T:

$$(T_h u_h)(\xi_{i,h}) = \sum_{j=1}^{l_h} \int_{G_{j,h}} K(\xi_{i,h}, y)\,dy\, u_h(\xi_{j,h}), \tag{5.29}$$

$$(T'_h u_h)(\xi_{i,h}) = \sum_{j=1}^{l_h} \mathscr{k}(\xi_{i,h}, \xi_{j,h}) \int_{G_{j,h}} \varkappa(\xi_{i,h}, y)\,dy\, u_h(\xi_{j,h}), \tag{5.29'}$$

$$(T''_h u_h)(\xi_{i,h}) = \sum_{\substack{j=1 \\ \mathrm{dist}(\xi_{i,h}, \mathrm{co}G_{j,h}) \geq h}}^{l_h} K(\xi_{i,h}, \xi_{j,h}) w_{j,h} u_h(\xi_{j,h}), \tag{5.29''}$$

$$(T'''_h u_h)(\xi_{i,h}) = \sum_{\substack{j=1 \\ j \neq i}}^{l_h} K(\xi_{i,h}, \xi_{j,h}) w_{j,h} u_h(\xi_{j,h}) + \beta_{i,h} u_h(\xi_{i,h}), \tag{5.29'''}$$

$$i = 1, \dots, l_h, \quad u_h \in E_h.$$

The operator $T \in \mathscr{L}(E,E)$ is compact (see Lemma 2.2); the operators T_h, T'_h, T''_h, $T'''_h \in \mathscr{L}(E_h, E_h)$ are also compact since E_h is finite dimensional. In Section 5.6 we prove that

$$T_h \dashrightarrow T \quad \text{compactly}; \tag{5.30}$$

in Sections 5.9–5.11 we prove that

$$\|T'_h - T_h\| \leq \text{const}\; \varepsilon^2_{\nu,h} \to 0 \quad \text{as} \quad h \to 0,$$

$$\|T''_h - T_h\| \leq \text{const}\; \varepsilon'_{\nu,h} \to 0 \quad \text{as} \quad h \to 0,$$

$$\|T'''_h - T_h\| \leq \text{const}\; \varepsilon'_{\nu,h} \to 0 \quad \text{as} \quad h \to 0$$

therefore we have also compact convergences $T'_h \dashrightarrow T$, $T''_h \dashrightarrow T$, $T'''_h \dashrightarrow T$. Thus, for methods (5.11)–(5.14), the conditions of Theorem 4.1 are fulfilled. Error estimate (4.12) yields for these methods, respectively,

$$\|u_h - p_h u\| \leq c_2 \|T_h p_h u - p_h T u\|,$$

$$\|u_h - p_h u\| \leq c_2 \|T'_h p_h u - p_h T u\| \leq c_2 \|T_h p_h u - p_h T u\| + c\,\varepsilon^2_{\nu,h}$$

$$\|u_h - p_h u\| \leq c_2 \|T''_h p_h u - p_h T u\| \leq c_2 \|T_h p_h u - p_h T u\| + c\,\varepsilon'_{\nu,h}$$

$$\|u_h - p_h u\| \leq c_2 \|T'''_h p_h u - p_h T u\| \leq c_2 (\|T_h p_h u - p_h T u\| + \|(T'''_h - T_h) p_h u\|)$$

where $u \in E = BC(G)$ is the solution of integral equation (3.1). Note that from condition (3.2), a similar inequality follows where ν is replaced by

μ $(\nu \le \mu < n)$. Recalling that $f \in C^{2,\mu}(G)$, we conclude with the help of Theorem 3.1 that $u \in C^{2,\mu}(G)$. Further, due to condition (5.16) and Lemma 2.3, $u = Tu + f$ satisfies a similar condition: for any $y^1, y^2 \in G$,

$$|u(y^1) - u(y^2)| \le \text{const} \cdot \left\{ \begin{array}{ll} d_G(y^1,y^2) & , \ \mu < n-1 \\ d_G(y^1,y^2)[1 + |\log d_G(y^1,y^2)|], & \mu = n-1 \\ [d_G(y^1,y^2)]^{n-\mu} & , \ \mu > n-1 \end{array} \right\}. \tag{5.31}$$

In Section 5.7 we prove, for any $u \in C^{2,\mu}(G)$ satisfying (5.31), we have

$$\|T_h p_h u - p_h T u\| \le \text{const}\ \varepsilon_{\nu,h}\varepsilon_{\mu,h}.$$

Hence, the error estimates take the following form:

for method (5.11) and (5.12), $\|u_h - p_h u\| \le \text{const}\ \varepsilon_{\nu,h}\varepsilon_{\mu,h}$;

for method (5.13), $\|u_h - p_h u\| \le \text{const}\ (\varepsilon_{\nu,h}\varepsilon_{\mu,h} + \varepsilon'_{\nu,h})$;

for method (5.14), $\|u_h - p_h u\| \le \text{const}\ (\varepsilon_{\nu,h}\varepsilon_{\mu,h} + \|(T_h''' - T_h)p_h u\|)$.

We recognize here estimate (5.17) for methods (5.11) and (5.12) as well estimate (5.22) for method (5.13). To obtain (5.17) for method (5.14), we prove in Section 5.11 that, for $u \in C^{2,\mu}(G)$ satisfying (5.31), we have

$$\|(T_h''' - T_h)p_h u\| \le \text{const}\ \varepsilon_{\nu,h}\varepsilon_{\mu,h}.$$

The error estimates for the prolonged functions (5.19), (5.21), (5.24) will be established by the way (see Sections 5.8 - 5.11).

5.6. Compact convergence of the discretized operators. Now we begin to fill the technical details of the proof of Theorem 5.1 outlined in Section 5.5. First we establish (5.30).

Lemma 5.2. Under conditions 1-3 of Theorem 5.1, $T_h \dashrightarrow T$ compactly (see (5.28) and (5.29) for the definitions of $T \in \mathcal{L}(E,E)$, $E = BC(G)$, and $T_h \in \mathcal{L}(E_h, E_h)$, $E_h = C(\Xi_h)$).

Proof. According to Lemma 4.1 and Definition 4.4 we have to prove that

$$\|T_h\| \le \text{const} \quad (0 < h < \bar{h}); \tag{5.32}$$

$$\|T_h p_h u - p_h T u\| \to 0 \quad \text{as} \quad h \to 0, \quad \forall u \in E = BC(G), \tag{5.33}$$

$$\|u_h\| \le 1 \ (0 < h < \bar{h}) \quad \Rightarrow \quad (T_h u_h) \text{ discretely compact.} \tag{5.34}$$

We have

$$\|T_h\| = \max_{1 \le i \le l_h} \sum_{j=1}^{l_h} \left| \int_{G_{j,h}} K(\xi_{i,h}, y)\, dy \right| \le \sup_{x \in G} \int_{G_h} |K(x,y)|\, dy$$

$$\to \sup_{x\in G} \int_G |K(x,y)|\,dy = \|T\| \quad \text{as} \quad h\to 0,$$

and (5.32) holds. Further, for any $u \in BC(G)$,

$$\|p_h Tu - T_h p_h u\| = \max_{1\le i\le l_h} \Big| \int_G K(\xi_{i,h}, y)\,u(y)\,dy - \sum_{j=1}^{l_h} \int_{G_{j,h}} K(\xi_{i,h}, y)\,dy\, u(\xi_{j,h})\Big|$$

$$\le \sup_{x\in G} \Big| \int_G K(x,y)\,u(y)\,dy - \sum_{j=1}^{l_h} \int_{G_{j,h}} K(x,y)\,u(\xi_{j,h})\,dy\Big|.$$

We approximate $G_{j,h}$ by $G'_{j,h}$ and G by $G'_h = \bigcup_{j=1}^{l_h} G'_{j,h} \subset G \cap G_h$ where $G'_{j,h} = G_{j,h}$ for inner cells and $G'_{j,h} \subset G_{j,h} \cap G$ has properties (5.6) and (5.7) for boundary incident cells. We have (see (5.5))

$$G\setminus G'_h \subset S_h, \qquad G_h \setminus G'_h \subset S_h$$

and, due to Lemma 5.1,

$$\sup_{x\in G} \int_{G\setminus G'_h} |K(x,y)|\,dy \le c\,\varepsilon_{\nu,h^2} \le c\,\varepsilon^2_{\nu,h},$$

$$\sup_{x\in G} \sum_{j=1}^{l_h} \int_{G_{j,h}\setminus G'_{j,h}} |K(x,y)|\,dy = \sup_{x\in G} \int_{G_h\setminus G'_h} |K(x,y)|\,dy \le c\,\varepsilon^2_{\nu,h}.$$

Thus,

$$\|p_h Tu - T_h p_h u\| \le \sup_{x\in G} \Big| \sum_{j=1}^{l_h} \int_{G'_{j,h}} K(x,y)[u(y) - u(\xi_{j,h})]\,dy \Big| \tag{5.35}$$

$$+ \text{const}\,\|u\|\,\varepsilon^2_{\nu,h}.$$

Now it is easy to obtain (5.33) noticing that $u \in BC(G)$ is uniformly continuous on sets $\{y\in G:\ \rho(y) \ge \delta\}$, $\delta > 0$, and due to Lemma 5.1

$$\sup_{x\in G} \int_{\{y\in G:\ \rho(y)<\delta\}} |K(x,y)|\,dy \le c\,\varepsilon_{\nu,\delta} \to 0 \quad \text{as} \quad \delta\to 0.$$

Let us prove (5.34). For given $u_h \in E_h$, $\|u_h\| \le 1$ $(0<h<\bar{h})$, define

$$u^{(h)} = \sum_{j=1}^{l_h} u_h(\xi_{j,h})\,\psi_{j,h} \in L^\infty(G)$$

where

$$\psi_{j,h}(x) = \begin{Bmatrix} 1, & x \in G\cap G_{j,h} \\ 0 & \text{otherwise} \end{Bmatrix}, \qquad j = 1, \dots, l_h.$$

Then $\|u^{(h)}\|_{L^\infty(G)} \le 1$ $(0<h<\bar{h})$, and due to Lemma 2.2, the family $(Tu^{(h)}) \subset BC(G)$ is relatively compact. Using (4.3) we see that the family $(p_h Tu^{(h)})$ is discretely compact. But $u_h = p_h u^{(h)}$ and repeating the argument by the proof of (5.35) we obtain

$$\|p_h Tu^{(h)} - T_h u_h\| = \|p_h Tu^{(h)} - Tp_h u^{(h)}\| \le \text{const}\, \varepsilon^2_{\nu,h} \to 0 \quad \text{as} \quad h \to 0$$

(note that $u^{(h)}(y) = u^{(h)}(\xi_{j,h})$ for $y \in G'_{j,h}$), therefore the family $(T_h u_h)$ is also discretely compact, and (5.34) holds.

5.7. Approximation error of the basic method. Lemma 5.3. Assume that conditions 1-3 of Theorem 5.1 are fulfilled and $\nu \le \mu < n$. Then, for any $u \in C^{2,\mu}(G)$ satisfying (5.31), we have

$$\|T_h p_h u - p_h Tu\|_{E_h} \le \text{const}\, \varepsilon_{\nu,h}\, \varepsilon_{\mu,h}$$

where $\varepsilon_{\nu,h}$ is given by (5.18); for definitions of $T \in \mathcal{L}(E,E)$, $E = BC(G)$, and $T_h \in \mathcal{L}(E_h, E_h)$, $E_h = C(\Xi_h)$, see (5.28) and (5.29).

Proof. We have $\varepsilon_{\nu,h} \le \varepsilon_{\mu,h}$. Due to (5.35), it suffices to establish that, for any $u \in C^{2,\mu}(G)$ satisfying (5.31),

$$\sup_{x\in G} \Big| \sum_{j=1}^{l_h} \int_{G_{j,h}} K(x,y)[u(y) - u(\xi_{j,h})]dy \Big| \le \text{const}\, \varepsilon_{\nu,h}\, \varepsilon_{\mu,h}. \tag{5.36}$$

(a) For $y \in G'_{j,h}$ we have $d_G(y, \xi_{j,h}) \le d_G\text{-}\operatorname{diam} G'_{j,h} \le h$ (see (5.4), (5.6) and (5.8)), and (5.31) yields

$$\max_{1\le j\le l_h} \sup_{y\in G'_{j,h}} |u(y) - u(\xi_{j,h})| \le \text{const}\, \varepsilon_{\mu,h}. \tag{5.37}$$

Introduce the index sets

$$J_h(x) = \{ j \in \mathbb{N}:\ 1 \le j \le l_h,\ \ \operatorname{dist}(x, \operatorname{co} G_{j,h}) < h \},$$

$$J_h(\partial G) = \{ j \in \mathbb{N}:\ 1 \le j \le l_h,\ \ \operatorname{dist}(\partial G, \operatorname{co} G_{j,h}) < h \}$$

where, for $A, B \subset \mathbb{R}^n$, $\operatorname{dist}(A,B) = \inf_{x\in A, y\in B} |x-y|$. Due (5.37) and (5.1),

$$\Big| \sum_{j\in J_h(\partial G)} \int_{G_{j,h}} K(x,y)[u(y) - u(\xi_{j,h})]dy \Big|$$

$$\le c\,\varepsilon_{\mu,h} \int_{\{y\in G:\ \rho(y)<2h\}} |K(x,y)|dy \le c'\varepsilon_{\mu,h}\, \varepsilon_{\nu,h}.$$

A similar inequality holds for the sum over $j \in J_h(x)$, too, since (see (3.3))

$$\int_{\{y\in G:\ |y-x|<2h\}} |K(x,y)|\,dy \le \text{const}\ \varepsilon_{\nu,h}.$$

The remaining part of the sum in (5.36) with $j \notin J_h(x)\cup J_h(\partial G)$ we estimate as follows:

$$\sup_{x\in G}\Big| \sum_{\substack{j=1\\ j\notin J_h(x)\cup J_h(\partial G)}}^{l_h} \int_{G_{j,h}} K(x,y)[u(y)-u(\xi_{j,h})]\,dy\Big| \le \lambda_h^1+\lambda_h^2$$

where

$$\lambda_h^1 = \sup_{x\in G} \sum_{\substack{j=1\\ j\notin J_h(x)\cup J_h(\partial G)}}^{l_h} \int_{G_{j,h}} |K(x,y)|\,|u(y)-u(\xi_{j,h})-u'(\xi_{j,h})(y-\xi_{j,h})|\,dy,$$

$$\lambda_h^2 = \sup_{x\in G} \sum_{\substack{j=1\\ j\notin J_h(x)\cup J_h(\partial G)}}^{l_h} \Big|\int_{G_{j,h}} K(x,y)\,u'(\xi_{j,h})(y-\xi_{j,h})\,dy\Big|$$

and $u'(x): \mathbb{R}^n \to \mathbb{R}$ (or $\mathbb{C}$) means the Fréchet derivative of u at x,

$$u'(x)y = \sum_{i=1}^{n} \frac{\partial u(x)}{\partial x_i}\, y_i \qquad (x\in G,\ y\in\mathbb{R}^n);$$

we took into account that $G'_{j,h} = G_{j,h}$ for $j\notin J_h(\partial G)$. To prove (5.36) we have to show that $\lambda_h^k \le \text{const}\ \varepsilon_{\mu,h}\,\varepsilon_{\nu,h}$ $(k=1,2)$.

(b) Let us estimate λ_h^1. For $y\in G_{j,h}$, $j\notin J_h(\partial G)$, we have

$$|u(y)-u(\xi_{j,h})-u'(\xi_{j,h})(y-\xi_{j,h})| \le \frac{1}{2}\sup_{0<t<1} |u''(ty+(1-t)\xi_{j,h})|\,|y-\xi_{j,h}|^2$$

where $u''(x): \mathbb{R}^n\times\mathbb{R}^n \to \mathbb{R}$ (or $\mathbb{C}$) is the second Fréchet derivative of u at x,

$$u''(x)yz = \sum_{i,j=1}^{n} \frac{\partial^2 u(x)}{\partial x_i \partial x_j}\, y_i z_j \qquad (x\in G,\quad y,z\in\mathbb{R}^n).$$

Note that for $y\in G_{j,h}$, $j\notin J_h(\partial G)$, $0<t<1$, we have

$$\frac{1}{2}\rho(y) \le \rho(ty+(1-t)\xi_{j,h}) \le 2\rho(y),$$

therefore the estimation of second derivatives of $u\in C^{2,\mu}(G)$ yields (see the definition of the space $C^{m,\nu}(G)$, Section 2.5)

$$|u''(ty+(1-t)\xi_{j,h})| \le \text{const}\left\{\begin{array}{ll} 1 & , \ \mu<n-2 \\ 1+|\log\rho(y)|, & \mu=n-2 \\ \rho(y)^{n-2-\mu} & , \ \mu>n-2 \end{array}\right\},$$

and

$$\lambda_h^1 \le ch^2 \sup_{x\in G} \int_{\{y\in G:\ \rho(y)>h\}} |K(x,y)| \left\{\begin{array}{ll} 1 & , \ \mu<n-2 \\ 1+|\log\rho(y)|, & \mu=n-2 \\ \rho(y)^{n-2-\mu} & , \ \mu>n-2 \end{array}\right\} dy.$$

In the case $\mu\le n-2$ we immediately obtain $\lambda_h^1 \le c'h^2 = c'\varepsilon_{\nu,h}\varepsilon_{\mu,h}$. Let us analyze the case $\mu>n-2$, $\nu>0$ (if (3.2) is satisfied with $\nu\le 0$ then we use a more coarse estimate with a $\nu>0$; recall that $\varepsilon_{\nu,h}=h$ for all $\nu<n-1$). Thus, $|K(x,y)| \le b|x-y|^{-\nu}$, $0<\nu<n$, and

$$\lambda_h^1 \le bch^2 \sup_{x\in G} \int_{\{y\in G:\ \rho(y)>h\}} |x-y|^{-\nu}\,\rho(y)^{n-2-\mu}\,dy$$

$$\le c'h^2 \sup_{x\in G} \sum_{k=1}^{[d/h]} \int_{\{y\in G:\ kh<\rho(y)<(k+1)h\}} |x-y|^{-\nu}\,dy\,(kh)^{n-2-\mu}$$

$$\le c''h^2\varepsilon_{\nu,h} \sum_{k=1}^{[d/h]} (kh)^{n-2-\mu} = c''h^{n-\mu}\,\varepsilon_{\nu,h} \sum_{k=1}^{[d/h]} k^{n-2-\mu} \le c'''\varepsilon_{\nu,h}\varepsilon_{\mu,h}$$

where $d=\operatorname{diam} G$ and $[d/h]$ is the integer part of d/h. We used inequality (5.1) for layers $\{y\in G:\ kh<\rho(y)<(k+1)h\}$ and took into account that

$$\sum_{k=1}^{[d/h]} k^{n-2-\mu} \le \int_1^{(d/h)+1} t^{n-2-\mu}\,dt \le \text{const}\left\{\begin{array}{ll} h^{-n+1+\mu} & , \ \mu<n-1 \\ 1+|\log h|, & \mu=n-1 \\ 1 & , \ \mu>n-1 \end{array}\right\}.$$

(c) Let us estimate λ_h^2. Due to the choice of the collocation points (see (5.8)) we have for $j\notin J_h(\partial G)$

$$\int_{G_{j,h}} (y-\xi_{j,h})\,dy = 0, \tag{5.38}$$

and λ_h^2 is representable in the form

$$\lambda_h^2 = \sup_{x\in G} \sum_{\substack{j=1 \\ j\notin J_h(x)\cup J_h(\partial G)}}^{l_h} \left| \int_{G_{j,h}} [K(x,y)-K(x,\xi_{j,h})]\,u'(\xi_{j,h})(y-\xi_{j,h})\,dy\right|.$$

For $y \in G_{j,h}$, $j \notin J_h(x)$, we have $\frac{1}{2}|x-y| \le |x-(ty+(1-t)\xi_{j,h})| \le 2|x-y|$ $(0<t<1)$ and, due to (3.2')

$$|K(x,y)-K(x,\xi_{j,h})| \le ch \sum_{i=1}^{n} \sup_{0<t<1} \left|\frac{\partial K(x,ty+(1-t)\xi_{j,h})}{\partial y_i}\right|$$

$$\le c'h \left\{\begin{matrix} 1 & , & \nu<-1 \\ 1+|\log|x-y||, & & \nu=-1 \\ |x-y|^{-\nu-1} & , & \nu>-1 \end{matrix}\right\} \tag{5.39}$$

Further, $\rho(y) \ge h$ for $y \in G_{j,h}$, $j \notin J_h(\partial G)$, and

$$|u'(\xi_{j,h})(y-\xi_{j,h})| \le ch|u'(\xi_{j,h})| \le c'h \left\{\begin{matrix} 1 & , & \mu=n-1 \\ 1+|\log\rho(y)|, & & \mu=n-1 \\ \rho(y)^{n-1-\mu} & , & \mu>n-1 \end{matrix}\right\}$$

$$\le c'h \left\{\begin{matrix} 1 & , & \mu<n-1 \\ 1+|\log h|, & & \mu=n-1 \\ h^{n-1-\mu} & , & \mu>n-1 \end{matrix}\right\} = c'\varepsilon_{\mu,h}.$$

We may assume that $\nu>-1$ using more coarse estimates of the kernel in the case $\nu \le -1$. Then (compare with (3.4))

$$\lambda_h^2 \le c\varepsilon_{\mu,h} h \int_{h<|x-y|<d} |x-y|^{-\nu-1} dy \le c'\varepsilon_{\mu,h} h \left\{\begin{matrix} 1 & , & \nu<n-1 \\ 1+|\log h|, & & \nu=n-1 \\ h^{n-\nu-1} & , & \nu>n-1 \end{matrix}\right\} = c'\varepsilon_{\mu,h}\varepsilon_{\nu,h}.$$

This completes the proof of Lemma 5.3.

5.8. Estimation of prolonged approximate solution. Together with Lemmas 5.2 and 5.3 we have established the main error estimate, (5.17), for PCCM (5.11). To finish the proof the assertion (i) of Theorem 5.1, it remains to establish estimate (5.20) for the prolonged function (5.19). We have

$$\sup_{x\in G} |u_h(x)-u(x)| = \sup_{x\in G} \left|\sum_{j=1}^{l_h} \int_{G_{j,h}} K(x,y)dy\, u_{j,h} - \int_G K(x,y)u(y)dy\right|$$

$$\le \sup_{x\in G} \sum_{j=1}^{l_h} \int_{G_{j,h}} |K(x,y)|dy\, |u_{j,h}-u(\xi_{j,h})|$$

$$+ \sup_{x\in G} \Big| \sum_{j=1}^{l_h} \int_{G_{j,h}} K(x,y)\,dy\, u(\xi_{j,h}) - \int_G K(x,y)u(y)\,dy \Big|$$

$$\le \sup_{x\in G} \int_{G_h} |K(x,y)|\,dy \max_{1\le j\le l_h} |u_{j,h} - u(\xi_{j,h})|$$

$$+ \sup_{x\in G} \Big| \sum_{j=1}^{l_h} \int_{G_{j,h}} K(x,y)[u(y) - u(\xi_{j,h})]\,dy \Big| + \mathrm{const}\,\|u\|\,\varepsilon^2_{\nu,h}$$

(see Section 5.6). Using (5.17) and (5.36) we obtain (5.20). Recall that, for the solution of (3.1), we have $u \in C^{2,\mu}(G)$ and (5.31), thus the use of (5.36) is legitimate.

As we see, the nature of the approximation $u_{j,h}$ is of no significance to obtain the accuracy (5.20) of the function $u_h(x)$ defined by (5.19). The only thing that we need here is the accuracy (5.17) of the approximation $u_{j,h}$ itself.

5.9. Proof of assertion (ii). Lemma 5.4. Let conditions 1 and 2 of Theorem 5.1 be fulfilled and let the kernel have the form $K(x,y) = \mathscr{k}(x,y)\varkappa(x,y)$ where $\mathscr{k} \in BC^2(G\times G)$ and $\varkappa$ satisfies (3.2) with $m = 2$. Then

$$\|T'_h - T_h\| \le \mathrm{const}\ \varepsilon^2_{\nu,h};$$

see (5.29) and (5.29') for the definitions of $T_h, T'_h \in \mathcal{L}(E_h, E_h)$, $E_h = C(\Xi_h)$.

Proof. We have

$$\|T_h - T'_h\| = \max_{1\le i\le l_h} \sum_{j=1}^{l_h} \Big| \int_{G_{j,h}} \mathscr{k}(\xi_{i,h},y)\varkappa(\xi_{i,h},y)\,dy - \mathscr{k}(\xi_{i,h},\xi_{j,h}) \int_{G_{j,h}} \varkappa(\xi_{i,h},y)\,dy \Big|$$

$$= \max_{1\le i\le l_h} \sum_{j=1}^{l_h} \Big| \int_{G_{j,h}} \varkappa(\xi_{i,h},y)[\mathscr{k}(\xi_{i,h},y) - \mathscr{k}(\xi_{i,h},\xi_{j,h})]\,dy \Big|$$

$$\le \sup_{x\in G} \sum_{j=1}^{l_h} \Big| \int_{G_{j,h}} \varkappa(x,y)[\mathscr{k}(x,y) - \mathscr{k}(x,\xi_{j,h})]\,dy \Big|$$

$$\le \sup_{x\in G} \sum_{j=1}^{l_h} \Big| \int_{G'_{j,h}} \varkappa(x,y)[\mathscr{k}(x,y) - \mathscr{k}(x,\xi_{j,h})]\,dy \Big| + c\varepsilon^2_{\nu,h}$$

(on the last step we used the argument of Section 5.6 again).

Now we are in a situation where we may apply an inequality of type (5.36) in which $\mu = \nu$, $\varkappa(x,y)$ stands for $K(x,y)$ and $\mathscr{k}(x,y)$ stands for $u(y)$.

Note that, due to the assumption $\mathcal{K}\in BC^2(G\times G)$, we have $\mathcal{K}(x,\cdot)\in C^{2,\nu}$ "uniformly with respect to x", i.e.

$$\sup_{x\in G}\|\mathcal{K}(x,\cdot)\|_{C^{2,\nu}(G)} \equiv \sup_{x\in G}\sum_{|\alpha|\le 2}\sup_{y\in G}\left(w_{|\alpha|-(n-\nu)}(y)\,|D_y^\alpha\mathcal{K}(x,y)|\right) < \infty .$$

Further, since $|\partial\mathcal{K}(x,y)/\partial y_i|$, $i=1,\dots,n$, are bounded, we have for any $y^1,y^2\in G$

$$\sup_{x\in G}|\mathcal{K}(x,y^1)-\mathcal{K}(x,y^2)| \le \text{const}\, d_G(y^1,y^2),$$

thus $\mathcal{K}$ satisfies (5.31) "uniformly with respect to x". The result of applying (5.36) is $\|T_h - T'_h\|\le\text{const}\,\varepsilon^2_{\nu,h}$. The Lemma is proved.

Turning back to Section 5.5 we now know that estimate (5.17) is true for method (5.12). The proof of estimate (5.20) for the prolonged function $u_h(x)$ defined in (5.21) contains no new ideas compared with Section 5.8 and the argument above and we propose it to reader as an exercise.

There is some space to weaken the conditions on $\mathcal{K}$ in Lemma 5.4 and Theorem 5.1 (ii).

5.10. Proof of assertion (iii). Lemma 5.5. Under conditions 1-3 of Theorem 5.1,

$$\|T''_h - T_h\| \le \text{const}\ \varepsilon'_{\nu,h};$$

see (5.29), (5.29") and (5.23) for the definitions.

Proof. The elements of the matrices $T_h=(t_{ij,h})$ and $T''_h=(t''_{ij,h})$, $i,j=1,\dots,l_h$, have the form

$$t_{ij,h} = \int_{G_{j,h}} K(\xi_{i,h},y)\,dy$$

$$t''_{ij,h} = \begin{cases} K(\xi_{i,h},\xi_{j,h})\,\text{meas}\, G_{j,h} & \text{if } \operatorname{dist}(\xi_{i,h},\text{co}G_{j,h})\ge h \\ 0 & \text{otherwise} \end{cases}.$$

We have

$$t_{ij,h}-t''_{ij,h} = \begin{cases} \int_{G_{j,h}} [K(\xi_{i,h},y)-K(\xi_{i,h},\xi_{j,h})]\,dy, & j\notin J_h(\xi_{i,h}) \\ \int_{G_{j,h}} K(\xi_{i,h},y)\,dy & , \ j\in J_h(\xi_{i,h}) \end{cases},\quad i,j=1,\dots,l_h,$$

$$\|T_h - T''_h\| = \max_{1\le i\le l_h}\sum_{j=1}^{l_h}|t_{ij,h}-t''_{ij,h}|$$

$$= \max_{1\le i\le l_h} \left\{ \sum_{j\in J_h(\xi_{i,h})} \left| \int_{G_{j,h}} K(\xi_{i,h},y)dy \right| + \sum_{\substack{j=1 \\ j\notin J_h(\xi_{i,h})}}^{l_h} \left| \int_{G_{j,h}} [K(\xi_{i,h},y) - K(\xi_{i,h},\xi_{j,h})]dy \right| \right\}$$

$$\le \lambda_h + \lambda'_h + \lambda''_h + c\varepsilon^2_{\nu,h}$$

where, with some designations of Sections 5.6 and 5.7,

$$\lambda_h = \sup_{x\in G} \sum_{j\in J_h(x)} \int_{G'_{j,h}} |K(x,y)|\,dy,$$

$$\lambda'_h = \sup_{x\in G} \sum_{j\in J_h(\partial G)\setminus J_h(x)} \int_{G'_{j,h}} |K(x,y) - K(x,\xi_{j,h})|\,dy,$$

$$\lambda''_h = \sup_{x\in G} \sum_{\substack{j=1 \\ j\notin J_h(\partial G)\cup J_h(x)}}^{l_h} \left| \int_{G_{j,h}} [K(x,y) - K(x,\xi_{j,h})]dy \right| .$$

With the help of (3.3) we find that

$$\lambda_h \le \sup_{x\in G} \int_{G\cap B(x,2h)} |K(x,y)|\,dy \le \text{const}\ \varepsilon'_{\nu,h}.$$

Further, using (5.39) with $\nu>-1$ (the case $\nu\le-1$ can be reduced again) and applying (5.1) if $\nu<n-1$, (5.2) if $\nu=n-1$ and (3.4) if $\nu>n-1$ we find

$$\lambda'_h \le ch \sup_{x\in G} \int_{\{y\in G:\, \rho(y)\le 2h,\ |x-y|>h\}} |x-y|^{-\nu-1}\,dy \le c'\varepsilon'_{\nu,h}.$$

Using (5.38) we rewrite

$$\lambda''_h = \sup_{x\in G} \sum_{\substack{j=1 \\ j\notin J_h(\partial G)\cup J_h(x)}}^{l_h} \left| \int_{G_{j,h}} [K(x,y) - K(x,\xi_{j,h}) - \frac{\partial K(x,\xi_{j,h})}{\partial y}(y-\xi_{j,h})]dy \right|$$

and estimate with the help of (3.2') and (3.4)

$$\lambda''_h \le ch^2 \sup_{x\in G} \int_{h<|x-y|<d} \left\{ \begin{array}{ll} 1 & ,\ \nu<-2 \\ 1+|\log|x-y||, & \nu=-2 \\ |x-y|^{-\nu-2} & ,\ \nu>-2 \end{array} \right\} dy \le c'\varepsilon'_{\nu,h}.$$

This completes the proof of Lemma 5.5.

Together with Lemma 5.5 we have established error estimate (5.22) for the approximation $u_{j,h}$ given by method (5.13); see Section 5.5 for the block argument. Now we prove estimate (5.25) for the prolonged $u_h(x)$ defined by (5.24). Denote by $u^o_{j,h}$ the solution of system (5.11) and (cf. (5.19))

$$u^o_h(x) = \sum_{j=1}^{l_h} \int_{G_{j,h}} K(x,y)\,dy\, u^o_{j,h} + f(x), \quad x \in G.$$

We have

$$u^o_h(x) - u_h(x) = \sum_{\substack{j=1\\ j \notin J_h(x)}}^{l_h} \int_{G_{j,h}} [K(x,y) - K(x,\xi_{j,h})]\,dy\, u^o_{j,h} + \sum_{j \in J_h(x)} \int_{G_{j,h}} K(x,y)\,dy\, u^o_{j,h}$$

$$+ \sum_{\substack{j=1\\ j \notin J_h(x)}}^{l_h} \int_{G_{j,h}} K(x,\xi_{j,h})\,dy\,(u^o_{j,h} - u_{j,h}).$$

Repeating the argument above and using (5.17) for $u^o_{j,h}$ and (5.22) for $u_{j,h}$ we find

$$\sup_{x \in G} |u^o_h(x) - u_h(x)| \le c\Big[\lambda_h + \lambda'_h + \lambda''_h + \varepsilon^2_{\nu,h} + \max_{1 \le i \le l_h}\big(|u^o_{j,h} - u(\xi_{j,h})| + |u_{j,h} - u(\xi_{j,h})|\big)\Big]$$

$$\le c'(\varepsilon'_{\nu,h} + \varepsilon^2_{\nu,h} + \varepsilon_{\nu,h}\varepsilon_{\mu,h}) \le c''(\varepsilon_{\nu,h}\varepsilon_{\mu,h} + \varepsilon'_{\nu,h}).$$

Now (5.25) for $u_h(x)$ follows from (5.20) for $u^o_h(x)$.

5.11. Proof of assertion (iv). Lemma 5.6. Assume that (5.26) and conditions 1–3 of Theorem 1 are fulfilled. Then

$$\|T'''_h - T_h\| \le \text{const}\, \varepsilon'_{\nu,h};$$

if in addition (5.27) holds then, for any function $u \in BC(G)$ satisfying (5.31) with $\nu \le \mu \le n$, we have

$$\|(T'''_h - T_h)p_h u\| \le \text{const}\, \varepsilon_{\nu,h}\varepsilon_{\mu,h}$$

(see (5.29) and (5.29''') for the definitions of $T_h, T'''_h \in \mathcal{L}(E_h, E_h)$, $E_h = C(\Xi_h)$).

Proof. For any $u_h \in E_h$ we have

$$((T_h - T'''_h)u_h)(\xi_{i,h}) = \sum_{\substack{j=1\\ j \ne i}}^{l_h} \int_{G_{j,h}} \big[K(\xi_{i,h},y) - K(\xi_{i,h},\xi_{j,h})\big]\,dy\,\big[u_h(\xi_{j,h}) - u_h(\xi_{i,h})\big]. \tag{5.40}$$

Thus,

$$\|T'''_h - T_h\| \le 2 \max_{1 \le i \le l_h} \sum_{\substack{j=1\\ j \ne i}}^{l_h} \int_{G_{j,h}} |K(\xi_{i,h},y) - K(\xi_{i,h},\xi_{j,h})|\,dy$$

$$\le 2\max_{1\le i\le l_h} \sum_{\substack{j=1\\ j\ne i,\ |\xi_{j,h}-\xi_{i,h}|\le 2h}}^{l_h} \Big[\int_{G_{j,h}} |K(\xi_{i,h},y)|\,dy + |K(\xi_{i,h},\xi_{j,h})|\,\mathrm{meas}\,G_{j,h}\Big]$$

$$+\ 2\max_{1\le i\le l_h} \sum_{\substack{j=1\\ |\xi_{j,h}-\xi_{i,h}|>2h}}^{l_h} \int_{G_{j,h}} |K(\xi_{i,h},y)-K(\xi_{i,h},\xi_{j,h})|\,dy \le \mathrm{const}\,\varepsilon'_{\nu,h},$$

since, due to (5.26) and (3.2), assuming $\nu>0$,

$$\sum_{\substack{j=1\\ j\ne i,\ |\xi_{j,h}-\xi_{i,h}|\le 2h}}^{l_h} |K(\xi_{i,h},\xi_{j,h})|\,\mathrm{meas}\,G_{j,h} \le b(2h)^{-\nu}\,\mathrm{meas}\,B(\xi_{i,h},3h) =$$

$$= ch^{n-\nu} \le c\varepsilon'_{\nu,h}$$

and other sums are always estimated by $c\varepsilon'_{\nu,h}$ in Section 5.10.

Take any $u\in BC(G)$ satisfying (5.31). Making use of (5.38) represent (5.40) in the form

$$((T_h-T_h''')p_h u)(\xi_{i,h}) = v_{i,h}+v'_{i,h}+v''_{i,h}+v'''_{i,h},\quad i=1,\dots,l_h,$$

where

$$v_{i,h} = \sum_{\substack{j=1\\ j\ne i,\ |\xi_{j,h}-\xi_{i,h}|\le 2h}}^{l_h} \int_{G_{j,h}} \big[K(\xi_{i,h},y)-K(\xi_{i,h},\xi_{j,h})\big]dy\,\big[u(\xi_{j,h})-u(\xi_{i,h})\big],$$

$$v'_{i,h} = \sum_{\substack{j=1\\ |\xi_{j,h}-\xi_{i,h}|>2h,\ j\in J_h(\partial G)}}^{l_h} \int_{G'_{j,h}} \big[K(\xi_{i,h},y)-K(\xi_{i,h},\xi_{j,h})\big]dy\,\big[u(\xi_{j,h})-u(\xi_{i,h})\big],$$

$$v''_{i,h} = \sum_{\substack{j=1\\ |\xi_{j,h}-\xi_{i,h}|>2h,\ j\notin J_h(\partial G)}}^{l_h} \int_{G_{j,h}} \Big[K(\xi_{i,h},y)-K(\xi_{i,h},\xi_{j,h})-\frac{\partial K(\xi_{i,h},\xi_{j,h})}{\partial y}(y-\xi_{j,h})\Big]dy$$

$$\cdot\big[u(\xi_{j,h})-u(\xi_{i,h})\big]$$

and $v'''_{i,h}$ are the errors caused by the approximation of $G_{j,h}$ by $G'_{j,h}\subset G_{j,h}\cap G$ satisfying (5.6) and (5.7). According to Lemma 5.1,

$$\max_{1\le i\le l_h} |v'''_{i,h}| \le c\varepsilon_{\nu,h^2} \le c\varepsilon^2_{\nu,h} \le c\varepsilon_{\nu,h}\varepsilon_{\mu,h}.$$

On the ground of (5.27), condition (5.31) takes the form

$$|u(y^1)-u(y^2)| \le \mathrm{const} \left\{ \begin{array}{ll} |y^1-y^2| & , \ \mu<n-1 \\ |y^1-y^2|(1+|\log|y^1-y^2||), & \mu=n-1 \\ |y^1-y^2|^{n-\mu} & , \ \mu>n-1 \end{array} \right\}, \quad y^1,y^2\in G.$$

For $|\xi_{j,h}-\xi_{i,h}| \le 2h$ this yields $|u(\xi_{j,h})-u(\xi_{i,h})| \le \mathrm{const}\,\varepsilon_{\mu,h}$, and estimating $|K(\xi_{i,h},y)-K(\xi_{i,h},\xi_{j,h})|$ by $|K(\xi_{i,h},y)|+|K(\xi_{i,h},\xi_{j,h})|$ we obtain

$$\max_{1\le i\le I_h} |v_{i,h}| \le c h^{n-\nu}\varepsilon_{\mu,h} \le c\varepsilon_{\nu,h}\varepsilon_{\mu,h}.$$

Since $\xi_{j,h}\in G'_{j,h}$ and d_G-diam$G'_{j,h} \le h$ (see (5.8) and (5.6)), any $y\in G'_{j,h}$ can be joined with $\xi_{j,h}$ by a polygonal path of the length $\frac{3}{2}h$ in $G\cap B(\xi_{j,h},\frac{3}{2}h)$. A consequence is that, for $y\in G'_{j,h}$, $|\xi_{i,h}-\xi_{j,h}|>2h$,

$$|K(\xi_{i,h},y)-K(\xi_{i,h},\xi_{j,h})| \le ch \sup_{z\in G\cap B(\xi_{j,h},\frac{3}{2}h)} |\xi_{i,h}-z|^{-\nu-1} \le c'h|\xi_{i,h}-y|^{-\nu-1}.$$

Further, for $y\in G'_{j,h}$, $|\xi_{i,h}-\xi_{j,h}|>2h$, inequality (5.31) yields

$$|u(\xi_{i,h})-u(\xi_{j,h})| \le \mathrm{const} \left\{ \begin{array}{ll} |\xi_{i,h}-y| & , \ \mu<n-1 \\ |\xi_{i,h}-y|(1+|\log|\xi_{i,h}-y||), & \mu=n-1 \\ |\xi_{i,h}-y|^{n-\mu} & , \ \mu>n-1 \end{array} \right\},$$

therefore

$$|v'_{i,h}| \le ch \int_{\{y\in G:\rho(y)<2h,\,|y-\xi_{i,h}|\ge h\}} \left\{ \begin{array}{ll} |y-\xi_{i,h}|^{-\nu} & , \ \mu<n-1 \\ |y-\xi_{i,h}|^{-\nu}(1+|\log|y-\xi_{i,h}||), & \mu=n-1 \\ |y-\xi_{i,h}|^{n-\mu-\nu-1} & , \ \mu>n-1 \end{array} \right\} dy.$$

With the help of (5.1), (5.2) and (3.4) we find that

$$\max_{1\le i\le I_h} |v'_{i,h}| \le c'h \left\{ \begin{array}{ll} h & , \ \mu<n-1 \\ h(1+|\log h|) & , \ \mu=n-1,\ \nu<n-1 \\ h(1+|\log h|)^2, & \mu=\nu=n-1 \\ h & , \ \mu>n-1,\ \nu+\mu<2(n-1) \\ h(1+|\log h|) & , \ \mu>n-1,\ \nu+\mu=2(n-1) \\ h^{2n-\nu-\mu-1} & , \ \mu>n-1,\ \nu+\mu>2(n-1) \end{array} \right\} \le c''\varepsilon_{\nu,h}\varepsilon_{\mu,h}.$$

Finally, for $y\in G_{j,h}$, $j\notin J_h(\partial G)$, $|\xi_{j,h}-\xi_{i,h}|>2h$ we have

$$\left|K(\xi_{i,h},y)-K(\xi_{i,h},\xi_{j,h})-\frac{\partial K(\xi_{i,h},\xi_{j,h})}{\partial y}(y-\xi_{j,h})\right| \le \mathrm{const}\, h^2\,|\xi_{i,h}-y|^{-\nu-2},$$

therefore

$$|v''_{i,h}| \le ch^2 \int_{\{y\in G:\,|y-\xi_{i,h}|\ge h\}} \begin{Bmatrix} |y-\xi_{i,h}|^{-\nu-1} & , \ \mu<n-1 \\ |y-\xi_{i,h}|^{-\nu-1}(1+|\log|y-\xi_{i,h}||), & \mu=n-1 \\ |y-\xi_{i,h}|^{n-\mu-\nu-2} & , \ \mu>n-1 \end{Bmatrix} dy$$

With the help of (3.4) we find that

$$\max_{1\le i\le l_h} |v''_{i,h}| \le c'h^2 \begin{Bmatrix} 1 & , \ \mu<n-1 \\ 1+|\log h| & , \ \mu=n-1,\ \nu<n-1 \\ (1+|\log h|)^2, & \mu=\nu=n-1 \\ 1 & , \ \mu>n-1,\ \nu+\mu<2(n-1) \\ 1+|\log h| & , \ \mu>n-1,\ \nu+\mu=2(n-1) \\ h^{2n-\nu-\mu-2} & , \ \mu>n-1,\ \nu+\mu>2(n-1) \end{Bmatrix} \le c''\varepsilon_{\nu,h}\varepsilon_{\mu,h}.$$

This completes the proof of the Lemma in the case $\nu>0$. The case $\nu\le 0$ is reducible to the case $\nu>0$ by making the estimate (3.2) more coarse; recall that $\varepsilon_{\nu,h}=h$ for all $\nu<n-1$.

Together with Lemma 5.6 we have established error estimate (5.17) for method (5.14). The proof of the error estimate for the prolonged approximation $u_h(x)$ is actually already given in Section 5.8.

We have completed the proof of Theorem 5.1.

5.12. Two grid iteration method. Systems (5.11)-(5.14) are of order $l_h \sim h^{-n}$. A direct solution of these large systems is possible only in the case of rough discretization, with a not too small discretization parameter h_*. Using two grid iteration methods, it is possible to solve these systems for much more fine discretizations with $h<<h_*$.

Introduce approximate partitions of G into cells $G_{j,h}$ $(j=1,\dots,l_h)$ and G_{j,h_*} $(j=1,\dots,l_{h_*})$ and choose corresponding collocation points $\xi_{j,h}\in G_{j,h}\cap G$ and $\xi_{j,h_*}\in G_{j,h_*}\cap G$ as in Section 5.2. For simplicity we assume that the following compatibility conditions are fulfilled:

(a) every cell $G_{j,h}$ $(j=1,\dots,l_h)$ is contained in a cell ("panel") G_{j',h_*} $(1\le j'\le l_{h_*})$ and conversely, every panel G_{j',h_*} $(j'=1,\dots,l_{h_*})$ is a union of some cells $G_{j,h}$ $(1\le j\le l_h)$;

(b) every collocation point ξ_{j',h_*} $(j'=1,\dots,l_{h_*})$ occurs as a collocation point for a cell $G_{j,h}\subset G_{j',h_*}$, i.e. $\Xi_{h_*}\subset\Xi_h$.

Introduce the space $E_{h_*}=C(\Xi_{h_*})$ and the connection operator $p_{h_*}\in\mathcal{L}(E,E_{h_*})$ in a similar way as $E_h=C(\Xi_h)$ and $p_h\in\mathcal{L}(E,E_h)$, $E=BC(G)$, see Section 4.2. Further, introduce the connection operators between E_h and E_{h_*} as follows:

$p_{h_*h}\in\mathcal{L}(E_h,E_{h_*})$, the restriction operator, $(p_{h_*h}u_h)(\xi_{j,h_*})=u_h(\xi_{j,h_*})$ for $u_h\in E_h$, $\xi_{j,h_*}\in\Xi_{h_*}$;

$p_{hh_*} \in \mathcal{L}(E_{h_*}, E_h)$, the piecewise constant prolongation operator,

$$(p_{hh_*} u_{h_*})(\xi_{j,h}) = u_{h_*}(\Pi_{h_*h}\xi_{j,h}) \quad \text{for} \quad u_{h_*} \in E_{h_*},\ \xi_{j,h} \in \Xi_h$$

where $\Pi_{h_*h}: \Xi_h \to \Xi_{h_*}$, $\Pi_{h_*h}\xi_{j,h} = \xi_{j',h_*}$ if $G_{j,h} \subset G_{j'h_*}$ (see conditions (a) and (b) above).

We treat systems (5.11)-(5.14) and their counterparts corresponding to the rough discretization, respectively, as equations

$$u_h = T_h u_h + p_h f \qquad \text{and} \qquad u_{h_*} = T_{h_*} u_{h_*} + p_{h_*} f,$$

$$u_h = T'_h u_h + p_h f \qquad \text{and} \qquad u_{h_*} = T'_{h_*} u_{h_*} + p_{h_*} f,$$

$$u_h = T''_h u_h + p_h f \qquad \text{and} \qquad u_{h_*} = T''_{h_*} u_{h_*} + p_{h_*} f,$$

$$u_h = T'''_h u_h + p_h f \qquad \text{and} \qquad u_{h_*} = T'''_{h_*} u_{h_*} + p_{h_*} f$$

where $T_h, T'_h, T''_h, T'''_h \in \mathcal{L}(E_h, E_h)$ are described in (5.29)-(5.29''') and T_{h_*}, T'_{h_*}, $T''_{h_*}, T'''_{h_*} \in \mathcal{L}(E_{h_*}, E_{h_*})$ have similar definitions, e.g.

$$(T_{h_*} u_{h_*})(\xi_{i,h_*}) = \sum_{j=1}^{l_{h_*}} \int_{G_{j,h_*}} K(\xi_{i,h_*}, y)\,dy\, u_{h_*}(\xi_{j,h_*}), \quad i = 1, \dots, l_{h_*},\ u_{h_*} \in E_{h_*}.$$

For system (5.11), the two grid iteration method has the form

$$\begin{aligned} &v_h^k = u_h^k - T_h u_h^k - p_h f \quad \text{(the residual of } u_h^k), \\ &u_h^{k+1} = u_h^k - v_h^k - p_{hh_*}(I_{h_*} - T_{h_*})^{-1} p_{h_*h} T_h v_h^k, \qquad k = 0,1,2,\dots . \end{aligned} \tag{5.41}$$

On the every iteration step one has to solve a system of type (5.11) but corresponding to the rough discretization. For system (5.12):

$$\begin{aligned} &v_h^k = u_h^k - T'_h u_h^k - p_h f, \\ &u_h^{k+1} = u_h^k - v_h^k - p_{hh_*}(I_{h_*} - T'_{h_*})^{-1} p_{h_*h} T'_h v_h^k, \quad k = 0,1,2,\dots; \end{aligned} \tag{5.42}$$

for system (5.13):

$$\begin{aligned} &v_h^k = u_h^k - T''_h u_h^k - p_h f, \\ &u_h^{k+1} = u_h^k - v_h^k - p_{hh_*}(I_{h_*} - T''_{h_*})^{-1} p_{h_*h} T''_h v_h^k, \quad k = 0,1,2,\dots; \end{aligned} \tag{5.43}$$

for system (5.14):

$$\begin{aligned} &v_h^k = u_h^k - T'''_h u_h^k - p_h f, \\ &u_h^{k+1} = u_h^k - v_h^k - p_{hh_*}(I_{h_*} - T'''_{h_*})^{-1} p_{h_*h} T'''_h v_h^k, \quad k = 0,1,2,\dots; \end{aligned} \tag{5.44}$$

For a convergence analysis, rewrite (5.41) in an equivalent form

$$u_h^{k+1} = T_{h,h_*} u_h^k + f_{h,h_*}, \qquad k=0,1,2,\dots \tag{5.45}$$

where

$$f_{h,h_*} = p_h f + p_{hh_*}(I_h - T_{h_*})^{-1} p_{h_* h} T_h p_h f \in E_h \tag{5.46}$$

$$\begin{aligned} T_{h,h_*} &= [I_h - p_{hh_*}(I_{h_*} - T_{h_*})^{-1} p_{h_* h}(I_h - T_h)]T_h \\ &= (I_h - p_{hh_*} p_{h_* h})T_h + p_{hh_*}(I_{h_*} - T_{h_*})^{-1}(p_{h_* h}T_h - T_{h_*} p_{h_* h})T_h \in \mathcal{L}(E_h, E_h). \end{aligned} \tag{5.47}$$

Iteration formulae (5.42)-(5.44) can be represented, respectively, in the form

$$u_h^{k+1} = T'_{h,h_*} u_h^k + f'_{h,h_*}, \quad u_h^{k+1} = T''_{h,h_*} u_h^k + f''_{h,h_*}, \quad u_h^{k+1} = T'''_{h,h_*} u_h^k + f'''_{h,h_*},$$
$$k=0,1,2,\dots,$$

where $f'_{h,h_*}, f''_{h,h_*}, f'''_{h,h_*} \in E_h$ and $T'_{h,h_*}, T''_{h,h_*}, T'''_{h,h_*} \in \mathcal{L}(E_h, E_h)$ are defined in a similar way as in (5.46) and (5.47) replacing T_h and T_{h_*} by T'_h and T'_{h_*} for mehod (5.42), by T''_h and T''_{h_*} for method (5.43) and by T'''_h and T'''_{h_*} for method (5.44).

Theorem 5.2. Let the following conditions be fulfilled:

1. $G \subset \mathbb{R}^n$ is open and bounded, ∂G satisfies condition (PS);
2. partitions of G and collocation points corresponding to the discretization parameters h and h_* ($h<h_*$) satsfy the conditions of Section 5.2 and compatibility conditions (a) and (b) introduced above;
3. kernel $K(x,y)$ satisfies condition (3.2) with $m=2$;
4. f satisfies (5.15) and (5.16);
5. integral equation (3.1) is uniquely solvable.

Then, for sufficiently small $h_*>0$, the following assertions are true.

(i) $\|T_{h,h_*}\| \le c\varepsilon_{\nu,h_*} < 1$ and, for iterations (5.41) and the exact solution u_h to system (5.11),

$$\|u_h^k - u_h\| \le \|u_h^0 - u_h\|(c\varepsilon_{\nu,h_*})^k, \qquad k=1,2,\dots, \tag{5.48}$$

where $\varepsilon_{\nu,h}$ is defined in (5.18) and the constant c is independent of h and h_*;

(ii) $\|T'_{h,h_*}\| \le c\varepsilon_{\nu,h_*} < 1$ and (5.48) holds for iterations (5.42) and the exact solution of system (5.12) provided that $\mathscr{k} \in BC^2(G\times G)$ and $\varkappa$ satisfies (3.2) with $m=2$;

(iii) $\|T''_{h,h_*}\| \le c\varepsilon_{\nu,h_*} < 1$ and (5.48) holds for iterations (5.43) and the exact solution of system (5.13);

(iv) $\|T'''_{h,h_*}\| \le c\varepsilon_{\nu,h_*} < 1$ and (5.48) holds for iterations (5.44) and the exact solution of system (5.14) provided that conditions (5.26) and (5.27) are satisfied.

<u>Proof</u>. Recall that $T_h \in \mathcal{L}(E_h,E_h)$ and $T \in \mathcal{L}(E,E)$ are compact and $T_h \dashrightarrow T$ compactly. Together with condition 5 of the Theorem this implies (see Lemma 4.2) that, for sufficiently small $h>0$, the operators $I_h - T_h$ are invertible and the inverses are uniformly bounded. Thus there are $h_1>0$ and $c_1 = \text{const}$ such that

$$\|(I_h - T_h)^{-1}\| \le c_1, \quad 0<h<h_1.$$

Note also that $\|p_{hh_*}\| = \|p_{h_*h}\| = 1$. To prove the inequality $\|T_{h,h_*}\| \le c\varepsilon_{\nu,h_*}$, it sufficies to establish that (see (5.47))

$$\|(I_h - p_{hh_*}p_{h_*h})T_h\| \le \text{const}\,\varepsilon_{\nu,h_*}, \tag{5.49}$$

$$\|(p_{h_*h}T_h - T_{h_*}p_{h_*h})T_h\| \le \text{const}\,\varepsilon_{\nu,h_*}. \tag{5.50}$$

Let us prove (5.49). To $u_h \in E_h$, let us coordinate a piecewise constant function $\bar{u}_h = \sum_{j=1}^{l_h} u_h(\xi_{j,h})\chi_{j,h}$ (see Section 5.3). We have $\|\bar{u}_h\|_{L^\infty(G_h)} = \|u_h\|_{E_h}$ and

$$(T_h u_h)(\xi_{i,h}) = \sum_{j=1}^{l_h} \int_{G_{j,h}} K(\xi_{i,h},y)\,dy\, u_h(\xi_{j,h}) = \int_{G_h} K(\xi_{i,h},y)\bar{u}_h(y)\,dy =$$

$$= \int_G K(\xi_{i,h},y)\bar{u}_h(y)\,dy + \int_{G_h\setminus G} K(\xi_{i,h},y)\bar{u}_h(y)\,dy - \int_{G\setminus G_h} K(\xi_{i,h},y)\bar{u}_h(y)\,dy\,.$$

According to Lemma 5.1

$$\int_{(G_h\setminus G)\cup(G\setminus G_h)} |K(x,y)|\,dy \le \text{const}\,\varepsilon_{\nu,h^2} \le \text{const}\,\varepsilon^2_{\nu,h}; \tag{5.51}$$

according to Lemma 2.3, for any $x^1,x^2 \in G$,

$$\left|\int_G [K(x^1,y)-K(x^2,y)]\bar{u}_h(y)\,dy\right| \le c\,\|\bar{u}_h\|_{L^\infty(G)} \begin{Bmatrix} d_G(x^1,x^2), & \nu<n-1 \\ d_G(x^1,x^2)(1+|\log d_G(x^1,x^2)|), & \nu=n-1 \\ d_G(x^1,x^2)^{n-\nu} & ,\ \nu>n-1 \end{Bmatrix}.$$

Therefore,

$$|(T_h u_h)(\xi_{i,h}) - (T_h u_h)(\xi_{k,h})| \le \text{const}\,\varepsilon^2_{\nu,h}\|u_h\|$$

$$+\ \text{const} \begin{Bmatrix} d_G(\xi_{i,h},\xi_{k,h}) & ,\ \nu<n-1 \\ d_G(\xi_{i,h},\xi_{k,h})(1+|\log d_G(\xi_{i,h},\xi_{k,h})|), & \nu=n-1 \\ d_G(\xi_{i,h},\xi_{k,h})^{n-\nu} & ,\ \nu>n-1 \end{Bmatrix} \|u_h\|.$$

Since $\xi_{i,h}$ and $\Pi_{h_*h}\xi_{i,h}$ are in a common panel G_{j',h_*}, we have $d_G(\xi_{i,h},\Pi_{h_*h}\xi_{i,h}) \le d_G\text{-diam}\,G_{j',h_*} \le h_*$ (see (5.6)), and we obtain

$$|(T_h u_h)\xi_{i,h}) - (T_h u_h)(\Pi_{h_*h}\xi_{i,h})| \le \text{const}\, \varepsilon_{\nu,h_*} \|u_h\|. \tag{5.52}$$

Note that, for $v_h = T_h u_h$,

$$(v_h - p_{hh_*} p_{h_*h} v_h)(\xi_{i,h}) = v_h(\xi_{i,h}) - v_h(\Pi_{h_*h}\xi_{i,h})$$

and, together with (5.52),

$$\|(I_h - p_{hh_*} p_{h_*h}) T_h u_h\| \le \text{const}\, \varepsilon_{\nu,h_*} \|u_h\|,$$

i.e. (5.49) holds.

Let us prove (5.50). Denoting again $v_h = T_h u_h$ we have

$$((p_{h_*h} T_h - T_{h_*} p_{h_*h}) T_h u_h)(\xi_{i',h_*}) = (T_h v_h^{*})(\xi_{i',h_*}) - (T_{h_*} p_{h_*h} v_h)(\xi_{i',h_*})$$

$$= \sum_{j=1}^{l_h} \int_{G_{j,h}} K(\xi_{i',h_*}, y)\,dy\, v_h(\xi_{j,h}) - \sum_{j'=1}^{l_{h_*}} \int_{G_{j',h_*}} K(\xi_{i',h_*}, y)\,dy\, v_h(\xi_{j',h_*})$$

$$= \sum_{j=1}^{l_h} \int_{G_{j,h}} K(\xi_{i',h_*}, y)\,dy\, [v_h(\xi_{j,h}) - v_h(\Pi_{h_*h}\xi_{j,h})].$$

Applying inequality (5.52) with $T_h u_h = v_h$ we obtain

$$\|(p_{h_*h} T_h - T_{h_*} p_{h_*h}) T_h u_h\| \le \sup_{x\in G} \int_{G_h} |K(x,y)|\,dy\, c\, \varepsilon_{\nu,h_*} \|u_h\| \le \text{const}\, \varepsilon_{\nu,h_*} \|u_h\|,$$

i.e. (5.50) holds. Thus, the estimate $\|T_{h,h_*}\| \le c\,\varepsilon_{\nu,h_*}$ is proved.

It is easy to check that the solution of the equation $u_h = T_h u_h + p_h f$ (the solution of system (5.11)) satisfies the equation

$$u_h = T_{h,h_*} u_h + f_{h,h_*}.$$

For $c\,\varepsilon_{\nu,h_*} < 1$, the last equation is uniquely solvable and iterations (5.45) converge to its solution with speed (5.48). In other words, iterations (5.41) converge to the solution of system (5.11) with speed (5.48). The assertion (i) of Theorem 5.2 is proved.

The assertions (ii)-(iv) of Theorem 5.2 now follow from the inequalities (see Lemmas 5.4 -5.6)

$$\|T'_h - T_h\| \le \text{const}\, \varepsilon^2_{\nu,h}, \quad \|T''_h - T_h\| \le \text{cons}\, \varepsilon'_{\nu,h}, \quad \|T'''_h - T_h\| \le \text{const}\, \varepsilon'_{\nu,h}$$

and the following joining result.

Let $\tilde{T}_h \in \mathcal{L}(E_h, E_h)$ be arbitrary approximations to $T_h \in \mathcal{L}(E_h, E_h)$ such that

$$\|\tilde{T}_h - T_h\| \le \text{const}\, \varepsilon_{\nu,h}, \quad 0 < h < \bar{h}. \tag{5.53}$$

Introduce the iterations

$$v_h^k = u_h^k - \tilde{T}_h u_h^k - p_h f,$$
$$u_h^{k+1} = u_h^k - v_h^k - p_{hh_*}(I_{h_*} - \tilde{T}_{h_*})^{-1} p_{h_* h} \tilde{T}_h v_h^k, \quad k = 0,1,2,\dots, \tag{5.54}$$

which can be represented in the form

$$u_h^{k+1} = \tilde{T}_{h,h_*} u_h^k + \tilde{f}_{h,h_*}, \quad k = 0,1,2,\dots,$$

where $\tilde{f}_{h,h_*} \in E_h$ and $\tilde{T}_{h,h_*} \in \mathcal{L}(E_h, E_h)$ are defined in a similar way as f_{h,h_*} in (5.46) and T_{h,h_*} in (5.47) replacing T_h and T_{h_*} by $\tilde{T}_h$ and $\tilde{T}_{h_*}$.

Lemma 5.7. Assume that (5.53) and conditions 1–5 of Theorem 5.2 are fulfilled. Then, for sufficiently small $h_* > 0$ and $0 < h < h_*$, we have

$$\|\tilde{T}_{h,h_*}\| \le c\varepsilon_{\nu,h_*} \tag{5.55}$$

and

$$\|u_h^k - u_h\| \le \|u_h^o - u_h\| (c\varepsilon_{\nu,h_*})^k, \quad k = 1,2,\dots \tag{5.56}$$

where u_h^k are defined by iteration method (5.54) and u_h is the (unique) solution of the equation $u_h = \tilde{T}_h u_h + p_h f$; constant c is independent of h and h_*.

Proof. It follows from (5.53) that

$$\|\tilde{T}_{h,h_*} - T_{h,h_*}\| \le \text{const}\, \varepsilon_{\nu,h_*}.$$

Thus, (5.55) is a consequence of the inequality $\|T_{h,h_*}\| \le \text{const}\, \varepsilon_{\nu,h_*}$ established by us proving Theorem 5.2 (i). Now (5.56) follows repeating the corresponding argument in the end part of the proof of Theorem 5.2 (i).

Lemma 5.7 is proved and the proof of Theorem 5.2 is completed.

5.13. Initial guesses. We propose two possible initial guesses for iteration method (5.41) which are based on the solution of system (5.11) on the rough discretization level.

Initial guess 1:

$$u_h^o = p_{hh_*} u_{h_*} = p_{hh_*}(I_{h_*} - T_{h_*})^{-1} p_{h_*} f; \tag{5.57}$$

we assert that, under conditions 1-5 of Theorem 5.2,

$$\|u_h^o - u_h\| \le \text{const}\, \varepsilon_{\mu,h_*} \tag{5.58}$$

where $u_h = (I_h - T_h)^{-1} p_h f$ is the solution of system (5.11). Indeed, applying (5.17) and (5.31) we find

$$\|u_h^o - u_h\| \le \|p_{hh_*}(u_{h_*} - p_{h_*} u)\| + \|p_{hh_*} p_{h_*} u - p_h u\| + \|p_h u - u_h\|$$

$$\le c(\varepsilon_{\nu,h_*}\varepsilon_{\mu,h_*}+\varepsilon_{\nu,h}\varepsilon_{\mu,h})+\sup_{\xi_{i,h}\in\Xi_h}|u(\xi_{i,h})-u(\Pi_{h_*h}\xi_{i,h})|\le c\,\varepsilon_{\mu,h_*}$$

where $u=(I-T)^{-1}f$ is the solution of integral equation (3.1).

Initial guess 2:

$$u_h^o = T_h p_{hh_*} u_{h_*} + p_h f = T_h p_{hh_*}(I_{h_*}-T_{h_*})^{-1} p_{h_*} f + p_h f; \tag{5.59}$$

we assert that, under conditions 1-5 of Theorem 5.2,

$$\|u_h^o - u_h\| \le \text{const}\, \varepsilon_{\nu,h_*}\varepsilon_{\mu,h_*}. \tag{5.60}$$

Indeed, introduce the prolonged approximation (5.19) on the rough discretization level:

$$\tilde{u}_{h_*}(x) = \sum_{j'=1}^{l_{h_*}} \int_{G_{j',h_*}} K(x,y)\,dy\, u_{h_*}(\xi_{j',h_*}) + f(x), \quad x\in G. \tag{5.61}$$

Estimate (5.20) provides

$$\sup_{x\in G} |\tilde{u}_{h_*}(x) - u(x)| \le c\,\varepsilon_{\nu,h_*}\varepsilon_{\mu,h_*}.$$

One can see from (5.59) and (5.61) that $\tilde{u}_{h_*}(\xi_{i,h}) = u_h^o(\xi_{i,h})$, $i=1,\dots,l_h$, therefore the last inequality yields

$$\|u_h^o - p_h u\| \le c\,\varepsilon_{\nu,h_*}\varepsilon_{\mu,h_*}.$$

On the other hand, due to (5.17),

$$\|u_h - p_h u\| \le c\,\varepsilon_{\nu,h}\varepsilon_{\mu,h}.$$

Now estimate (5.60) follows from these two inequalities.

Similar initial guesses may be used in methods (5.42) - (5.44). Formally, in (5.57) and (5.59), the operators T_h and T_{h_*} must be replaced by T_h' and T_{h_*}' (for method (5.42)), by T_h'' and T_{h_*}'' (for method (5.43)) or by T_h''' and T_{h_*}''' (for method (5.44)).

5.14. Amount of arithmetical work. In practical computations one usually stops the itereations as the residual is less than a given treshold. Let us analyze the asymptotical number of iteration steps as $h\to 0$. This number depends on the relation between h and h_* as well on the accuracy treshold. There is no motivation to compute u_h^k much more accurately than $\|u_h^k - u_h\| \le \varepsilon_{\nu,h}\varepsilon_{\mu,h}$ since method (5.11) itself is of the accuracy $c\,\varepsilon_{\nu,h}\varepsilon_{\mu,h}$, see Theorem 5.1. In the sequal we assume that conditions 1-5 of Theorem 5.2 are fulfilled whereby, for simplicity, $\mu=\nu$.

Beginning with initial guess 1 we obtain the accuracy

$$\|u_h^k - u_h\| \le \varepsilon_{\nu,h}^p, \quad p\ge 2, \tag{5.62}$$

of iterations (5.41) provided that $(c\varepsilon_{\nu,h_*})^{k+1} \le \varepsilon^{p}_{\nu,h}$ (see (5.48) and (5.58)), i.e.

$$k \ge \frac{p\,|\log \varepsilon_{\nu,h}|}{|\log \varepsilon_{\nu,h_*} + \log c|} - 1$$

(it is natural to assume that $\varepsilon_{\nu,h}<1$, $c\varepsilon_{\nu,h_*}<1$). Note that

$$\log \varepsilon_{\nu,h} \sim \begin{cases} \log h \quad , & \nu \le n-1 \\ (n-\nu)\log h, & \nu > n-1 \end{cases} \quad \text{as} \quad h \to 0.$$

Choosing $h_* \sim h^{\tau}$ $(0<\tau<1)$ we see that accuracy (5.62) will be achieved asymptotically in $k=\{p/\tau\}-1$ iteration steps where $\{q\}$ is the smallest integer exceeding q. For initial guess 2, this number is $k=\{p/\tau\}-2$. Remarkable is that these numbers do not depend on h and h_* separately but only on the relation $h_* \sim h^{\tau}$, $0<\tau<1$. The following Table illustrates the asymptotical number of iterations to obtain the accuracy $\|u^k_h - u_h\| = \mathcal{O}(\varepsilon^2_{\nu,h})$ for some strategies $h_* \sim h^{\tau}$.

Strategy	$h_* \sim h^{1/2}$	$h_* \sim h^{1/3}$
Initial guess 1	$k=4$	$k=6$
Initial guess 2	$k=3$	$k=5$

Table 5.1

Now let us analyze what an iteration step costs. The most extensive computational work during an iteration step (5.41) is caused by the term $T_h u^k_h$ — the cost is l^2_h additions and multiplications. The computation of a term of the form $w_{h_*} = (I_{h_*} - T_{h_*})^{-1} v_{h_*}$ can be realized as solving the system $w_{h_*} = T_{h_*} w_{h_*} + v_{h_*}$; using the Gauss algorithm it costs $l^3_{h_*}/3$ additions and multiplications. Putting $h_* \sim h^{\tau}$, $0<\tau<2/3$, and taking into account (or assuming) that then $l_{h_*} \sim l^{\tau}_h$, the use of Gauss algorithm for the computation of w_{h_*} will be cheeper than the evaluation of $T_h u^k_h$. Note that the computation of the terms $p_{h_* h} T_h v^k_h$ or $T_h p_{h h_*} u_{h_*}$ (the last one

Strategy	$h_* \sim h^{1/2}$	$h_* \sim h^{1/3}$
Initial guess 1	$4 l^2_h$ additions and multiplications	$6 l^2_h$ additions and multiplications
Initial guess 2	$3 l^2_h$ additions and multiplications	$5 l^2_h$ additions and multiplications

Table 5.2.

occurs in the initial guess) costs $l_h l_{h_*}$ additions and multiplications, and other operators in (5.41) are of $\mathcal{O}(l_h)$ additions and multiplications. Table 5.2 illustrates the asymptotical amount of the arithmetical work to obtain the accuracy $\|u_h^k - u_h\| = o(\varepsilon^2_{\nu,h})$ for some strategies $h_* \sim h^\tau$. For iteration methods (5.42)-(5.44), the estimation of the total arithmetic work is similar.

In the case of convolution type kernels

$$K(x,y) = a(x)\varkappa(x-y)b(y),$$

using the regular partitions of G described in Section 5.2 (iii) with collocation points (5.10), an evaluation of $T_h'' v_h$ and $T_h''' v_h$ can be performed in $\mathcal{O}(l_h \log_2 l_h)$ arithmetical operations using multidimensional fast Fourier transformations (FFT). Correspondingly, systems (5.13) and (5.14) can be solved with the accuracy $o(\varepsilon^2_{\nu,h})$ in $\mathcal{O}(l_h \log_2 l_h)$ arithmetical operations applying iteration methods (5.43) and (5.44) with $h_* \sim h^{1/3}$.

Let us explain where the FFT can be applied computing

$$(T_h'' v_h)(\xi_{\lambda,h})$$

$$= \sum_{\{\lambda' \in \mathbb{Z}^n :\, \xi_{\lambda',h} \in G,\, \lambda' \neq \lambda\}} a(\xi_{\lambda,h})\varkappa((\lambda-\lambda')h)b(\xi_{\lambda',h}) \operatorname{meas} G_{\lambda',h} v_h(\xi_{\lambda',h}), \quad \xi_{\lambda,h} \in \Xi_h$$

(to simplity the writing we took $h_1 = \ldots = h_n = h$). We have here three pointwise multiplications of grid functions (each of them costs l_h arithmetical operations) and a discrete convolution

$$\sum_{\{\lambda' \in \mathbb{Z}^n :\, \xi_{\lambda',h} \in G,\, \lambda' \neq \lambda\}} \varkappa((\lambda-\lambda')h)\, w_h(\lambda' h), \qquad \xi_{\lambda,h} \in \Xi_h$$

which can be performed in $\mathcal{O}(l_h \log_2 l_h)$ arithmetical operations using the multidimensional FFT (see e.g. Ivanov (1986)) precedently extending the sum up to a cube $\Omega \supset G$ defining $w_h(\lambda' h) = 0$ outside of G.

The operator T_h''' differs from T_h'' only by the diagonal elements of the matrices; an applying of a diagonal matrix to v_h costs l_h arithmetical operations. So, $T_h''' v_h$ can also be computed in $\mathcal{O}(l_h \log_2 l_h)$ arithmetical operations.

5.15. Eigenvalue problem. Recall that the operators $T \in \mathcal{L}(E,E)$ and T_h, $T_h', T_h'', T_h''' \in \mathcal{L}(E_h,E_h)$ are compact and $T_h \dashrightarrow T$, $T_h' \dashrightarrow T$, $T_h'' \dashrightarrow T$, $T_h''' \dashrightarrow T$ compactly (see Lemmas 5.2, 5.4, 5.5 and 5.6 where the needed assumptions are formulated). This enables to apply Theorem 4.2 for the discretizations under consideration. Detailed formulations are rather close to that of Theorem 4.2. We note here only that, in error estimates (4.15) and (4.16),

$$e_h \le \mathrm{const}\,\varepsilon^2_{\nu,h} \qquad \text{for } T_h, T'_h \text{ and } T'''_h,$$

$$e_h \le \mathrm{const}\,\varepsilon'_{\nu,h} \qquad \text{for } T''_h$$

with $\varepsilon_{\nu,h}$ and $\varepsilon'_{\nu,h}$ defined in (5.18) and (5.23). Indeed, $W(\lambda_0,T) \subset C^{2,\nu}(G)$ due to Remark 3.2 and any $u_0 \in W(\lambda_0,T)$ satisfies (5.31) due to Lemma 2.3, and the estimates of e_h follows from Lemmas 5.3-5.6.

5.16. Exercises. 5.16.1. The condition (PS) introduced in Section 5.1 does not imply the compactness of G^* (see Section 2.1 for the definition of G^*). Discuss the following example (see Figure 5.6): the boundary of $G \subset \mathbb{R}^2$ is given by 4 smooth curves

$$x_2 = \pm x_1^3 \quad (0 \le x_1 \le 1), \qquad x_1 = 1 \quad (-1 \le x_2 \le 1), \qquad x_2 = x_1^3 \sin(1/x_1) \quad (0 < x_1 \le 1).$$

Thus, G consists of a countable number of connectivity components and G^* is non-compact. Modify the example introducing a third dimension so that $G \subset \mathbb{R}^3$ is connected, ∂G satisfies (PS) and G^* remains non-compact.

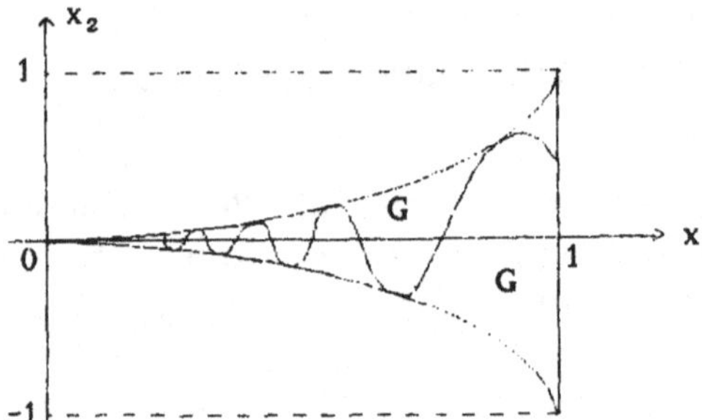

Fig. 5.6.

5.16.2. In general, there does not exist exact partitions of G such that $d_G\text{-}\mathrm{diam}\, G_{j,h} \le h$, $0 < h < \bar{h}$ (see exercise 2.7.1). Prove that for any open bounded set $G \subset \mathbb{R}^n$ there exist approximate partitions of G satisfying the conditions of Section 5.2.

5.16.3. Prove that Theorem 5.1 remains true if instead of the centroids $\xi_{j,h}$ of $G_{j,h}$ with $\mathrm{co}\, G_{j,h} \subset G$ their approximations $\xi'_{j,h}$ of the accuracy $|\xi'_{j,h} - \xi_{j,h}| \le ch^2$ are used.

5.16.4. Construct a modification of method (5.14) where

$$\int_{G_h} K(\xi_{i,h}, y)\,dy = \sum_{j=1}^{l_h} \int_{G_{j,h}} K(\xi_{i,h}, y)\,dy$$

is evaluated as in method (5.12). Prove that under conditions introduced in assertions (iv) and (ii) of Theorem 5.1 the accuracy (5.17) is maintained.

5.16.5. Prove estimate (5.20) for the prolonged function $u_h(x)$ defined in (5.21).

5.16.6. Analyzing the proof of Lemma 5.4, formulate more general conditions on k under which Theorem 5.1 (ii) remains true.

5.16.7. Establish error bounds the initial guesses for iteration methods (5.42), (5.43) and (5.44) described in the end of Section 5.13.

5.16.8. Give extended formulations of the results concerning the approximation of the eigenvalue problem.

5.16.9. In some cases the solution to integral equation (3.1) may be more regular than $u \in C^{2,\mu}(G)$ under the conditions of Theorem 5.1. Assume that $u \in BC^2(G)$ and sharp partitions of G are used (i.e. $\overline{G}_h = \overline{G}$). Prove the following refinement of estimate (5.17) for method (5.11):

$$\max_{1 \le i \le l_h} |u_{i,h} - u(\xi_{i,h})| \le \text{const}\, h\, \varepsilon_{\nu,h}.$$

Establish error estimates for methods (5.12), (5.13) and (5.14), too. (The situation considered here is interesting due to the analogy to boundary integral equation (0.3) on a smooth boundary ∂G, with $f \in C^2(\partial G)$ and a smooth weakly singular kernel $K(x,y)$, $x,y \in \partial G$, — then $u \in C^2(\partial G)$.)

6. COMPOSITE CUBATURE ALGORITHMS

As we saw in Chapter 5, the simplest cubature formula method loses the accuracy of the basic PCCM if the singularity of the kernel of equation (3.1) is of the order $\nu \geq n-2$. In this Chapter we propose composite cubature algorithms for the evaluation of the coefficients of the system. Our purpose is to restore the accuracy of the basic PCCM consuming to every coefficient of the system in the average only a finite number of arithmetical operations not depending on h.

6.1. Partition of G and composite cubature formulae. In Chapter 5 we used possibly approximate partitions of G with exact values of meas $G_{\lambda,h}$. In this Chapter we, conversely, start from sharp partitions of G but allow approximate values of $\operatorname{meas} G_{\lambda,h}$. To some extent, both approaches are equivalent.

Let us denote H_C the collection of $h=(h_1,\dots,h_n)\in\mathbb{R}^n$ with $h_i>0$, $|h|/h_i \leq c$ $(i=1,\dots,n)$, $|h|=(h_1^2+\dots+h_n^2)^{1/2}$. To any $h\in H_C$, we coordinate a regular partitions of $\mathbb{R}^n$ into the boxes

$$B_{\lambda,h} = \{x\in\mathbb{R}^n:\ \lambda_i h_i < x_i < (\lambda_i+1)h_i,\ i=1,\dots,n\},\ \lambda=(\lambda_1,\dots,\lambda_n)\in\mathbb{Z}^n;$$

the center of the box $B_{\lambda,h}$ is given by

$$\xi^0_{\lambda,h} = ((\lambda_1+\tfrac{1}{2})h_1,\dots,(\lambda_n+\tfrac{1}{2})h_n),\quad \lambda\in\mathbb{Z}^n.$$

Introduce a partition of G into sets $G_{\lambda,h}$, $\lambda\in\Lambda_h\subset\mathbb{Z}^n$,

$$G_{\lambda,h}\cap G_{\lambda',h} = \emptyset \quad \text{for} \quad \lambda\neq\lambda', \qquad \bigcup_{\lambda\in\Lambda_h}\overline{G}_{\lambda,h} = \overline{G}, \tag{6.1}$$

which is coordinated to the partition of $\mathbb{R}^n$:

$$\{\lambda\in\mathbb{Z}^n:\ \xi^0_{\lambda,h}\in G\}\subseteq\Lambda_h\subseteq\{\lambda\in\mathbb{Z}^n: G\cap B_{\lambda,h}\neq\emptyset\}, \tag{6.2}$$

$$G_{\lambda,h} = B_{\lambda,h} \ \text{ if } \ \xi^0_{\lambda,h}\in G,\ \ \operatorname{dist}(\xi^0_{\lambda,h},\partial G)>|h|, \tag{6.3}$$

$$\partial G_{\lambda,h}\subset \bigcup_{\lambda'}\partial B_{\lambda',h}\cup\partial G,\quad \lambda\in\Lambda_h. \tag{6.4}$$

We assume also that

$$d_G\text{-}\operatorname{diam} G_{\lambda,h} \le \text{const}\,|h|, \quad \lambda \in \Lambda_h, \tag{6.5}$$

where the constant is independent of $h \in H_C$. Note that (6.5) can be viewed as a smoothness condition to ∂G, since for inner cells $G_{\lambda,h} = B_{\lambda,h}$ we have $d_G\text{-}\operatorname{diam} G_{\lambda,h} = \operatorname{diam} B_{\lambda,h} = |h|$. Condition (6.4) means that near ∂G the cells $G_{\lambda,h}$ are constructed as unions of some blocks $G \cap B_{\lambda',h}$ (see Figures 6.1, 6.2) or of connectivity components of those (the case of inner boundary).

In every $G_{\lambda,h}$, $\lambda \in \Lambda_h$, we choose a node $\xi_{\lambda,h}$ as follows:

$$\begin{aligned} &\xi_{\lambda,h} = \xi^0_{\lambda,h} \quad \text{if} \quad G_{\lambda,h} = B_{\lambda,h}, \\ &\text{otherwise } \xi_{\lambda,h} \in G_{\lambda,h} \text{ is fixed arbitrarily.} \end{aligned} \tag{6.6}$$

Let $w_{\lambda,h}$, $\lambda \in \Lambda_h$, be positive weights such that

$$\begin{aligned} &w_{\lambda,h} = \operatorname{meas} B_{\lambda,h} = h_1 \cdot \ldots \cdot h_n \quad \text{if} \quad G_{\lambda,h} = B_{\lambda,h}, \\ &\text{otherwise } |w_{\lambda,h} - \operatorname{meas} G_{\lambda,h}| \le \text{const}\,|h|\, h_1 \cdot \ldots \cdot h_n. \end{aligned} \tag{6.7}$$

We shall use the cubature formula

$$\int_{G_{\lambda,h}} v(y)\,dy \approx v(\xi_{\lambda,h})\, w_{\lambda,h}, \quad \lambda \in \Lambda_h, \tag{6.8}$$

and its composite versions

$$\int_{G_{\lambda,h}} v(y)\,dy \approx \sum_{\{\mu \in \Lambda_{N^{-1}h}:\, G_{\mu,N^{-1}h} \subset G_{\lambda,h}\}} v(\xi_{\mu,N^{-1}h})\, w_{\mu,N^{-1}h}$$

where $N \ge 1$ is an integer. Thereby we assume that $G_{\mu,N^{-1}h} \subset G_{\lambda,h}$ if $G_{\mu,N^{-1}h} \cap G_{\lambda,h}$ is non-void.

Let us present some examples of partitions of G satisfying (6.1) - (6.4).

Example 6.1. $G_{\lambda,h} = G \cap B_{\lambda,h}$, $\lambda \in \Lambda_h = \{\lambda \in \mathbb{Z}^n:\ G \cap B_{\lambda,h} \ne \emptyset\}$, see Figure 6.1. One may redefine Λ_h and $G_{\lambda,h}$ joining smaller cells to their neighbours as it is done on Figure 6.2.

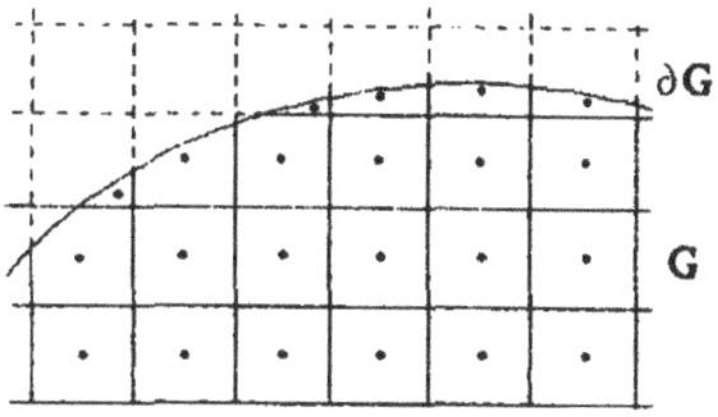

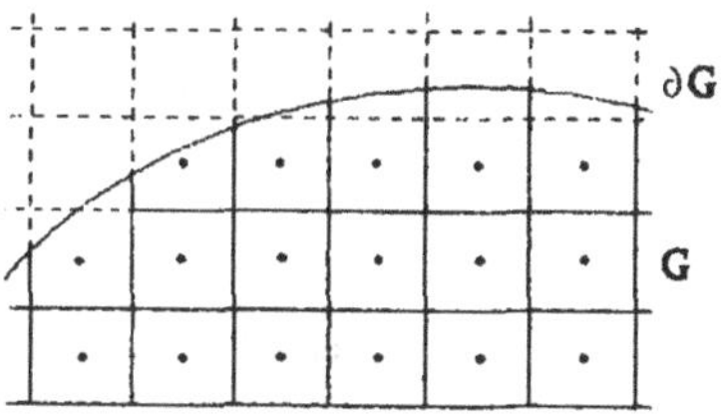

Fig.6.1. Fig.6.2.

Example 6.2. Let $\Lambda_h = \{\lambda \in \mathbb{Z}^n:\ \xi^0_{\lambda,h} \in G\}$. On the first step we define $G_{\lambda,h}$, $\lambda \in \Lambda_h$, as the connectivity component of $G \cap B_{\lambda,h}$ containing $\xi^0_{\lambda,h}$. On the second step we expand the boundary-incident cells joining to them the remaining parts of G (see Figure 6.2). In this example it is possible to choose $\xi_{\lambda,h}=\xi^0_{\lambda,h}$ for all $\lambda\in\Lambda_h$.

If ∂G is C^2-smooth inside a box $B_{\lambda,h}$ then $w_{\lambda,h}$ satisfying (6.7) can be found e.g. as meas $\tilde{G}_{\lambda,h}$ where $\tilde{G}_{\lambda,h}$ is obtained from $G_{\lambda,h}$ approximating ∂G by a tangent or secant plain. This procedure of the finding of a weight $w_{\lambda,h}$ costs $O(1)$ arithmetical operations.

6.2. Basic methods. In the designations of Section 6.1 the basic piecewise constant collocation method (PCCM) to solve integral equation (3.1) looks as

$$u_{\lambda,h}=\sum_{\lambda'\in\Lambda_h}\int_{G_{\lambda',h}}K(\xi_{\lambda,h},y)\,dy\,u_{\lambda',h}+f(\xi_{\lambda,h}),\qquad \lambda\in\Lambda_h \tag{6.9}$$

(cf. (5.11)). Consider the cubature formula method (CFM) corresponding to cubature formula (6.8), too:

$$u_{\lambda,h}=\sum_{\{\lambda'\in\Lambda_h:\,|\xi_{\lambda,h}-\xi_{\lambda',h}|\ge c_1|h|\}}K(\xi_{\lambda,h},\xi_{\lambda',h})\,w_{\lambda',h}u_{\lambda',h}+f(\xi_{\lambda,h}),\quad \lambda\in\Lambda_h \tag{6.10}$$

(cf.(5.13)). Theorem 5.1 (i), (iii) can be reformulated as follows.

Theorem 6.1. Assume that the following conditions are fulfilled:

1. $G\subset\mathbb{R}^n$ is open and bounded, ∂G satisfies condition (PS);
2. conditions (6.1)-(6.7) concerning the partition of G are satisfied;
3. the kernel $K(x,y)$ satisfies (3.2) with $m=2$, $\nu<n$;
4. $f\in C^{2,\nu}(G)$ and satisfies (5.16) with $\mu=\nu$;
5. equation (3.1) has a unique solution $u\in BC(G)$.

Then there exists a $\delta_0>0$ such that, for all $h\in H_C$ with $|h|<\delta_0$, systems (6.9) and (6.10) are uniquely solvable and the following error estimates hold:

(a) for PCCM (6.9),

$$\max_{\lambda\in\Lambda_h}|u_{\lambda,h}-u(\xi_{\lambda,h})|\le\text{const}\,(\varepsilon_{\nu,h})^2, \tag{6.11}$$

$$\sup_{x\in G}|u_h(x)-u(x)|\le\text{const}\,(\varepsilon_{\nu,h})^2 \tag{6.12}$$

where

$$u_h(x)=\sum_{\lambda'\in\Lambda_h}\int_{G_{\lambda',h}}K(x,y)\,dy\,u_{\lambda',h}+f(x),\qquad x\in G, \tag{6.13}$$

$$\varepsilon_{\nu,h}=\left\{\begin{array}{ll}|h| & ,\ \nu<n-1\\ |h|\,(1+|\log|h||), & \nu=n-1\\ |h|^{n-\nu} & ,\ \nu>n-1\end{array}\right\}; \tag{6.14}$$

(b) for CFM (6.10),

$$\max_{\lambda\in\Lambda_h} |u_{\lambda,h} - u(\xi_{\lambda,h})| \le \text{const}\, \varepsilon'_{\nu,h}, \tag{6.15}$$

$$\sup_{x\in G} |u_h(x) - u(x)| \le \text{const}\, \varepsilon'_{\nu,h} \tag{6.16}$$

where

$$u_h(x) = \sum_{\{\lambda'\in\Lambda_h:\, \operatorname{dist}(x, G_{\lambda',h}) \ge c_1|h|\}} K(x, \xi_{\lambda',h})\, w_{\lambda',h} u_{\lambda',h} + f(x), \quad x\in G, \tag{6.17}$$

$$\varepsilon'_{\nu,h} = \left\{ \begin{array}{ll} |h|^2 & , \ \nu < n-2 \\ |h|^2(1+|\log|h||), & \nu = n-2 \\ |h|^{n-\nu} & , \ \nu > n-2 \end{array} \right\}. \tag{6.18}$$

For $\nu<n-2$, CFM achieves the accuracy of the basic PCCM; for $\nu\ge n-2$, CFM loses in accuracy. Our purpose is to refine the cubature evaluation of the coefficients

$$t_{\lambda,\lambda',h} = \int_{G_{\lambda',h}} K(\xi_{\lambda,h}, y)\,dy, \quad \lambda,\lambda'\in\Lambda_h, \tag{6.19}$$

so that the accuracy (6.11) will be restored and that only $\mathcal{O}(l_h^2)$ arithmetical operations will be needed to evaluate all l_h^2 integrals (6.19). Here $l_h = \operatorname{card}\Lambda_h$ is the number of unknowns of system (6.9). Due to condition (6.2), l_h is of the order $\mathcal{O}(|h|^{-n})$.

6.3. Refined cubatures for the evaluation of the coefficients. We shall use cubature formula (6.8) and its composite version:

$$\tilde{t}_{\lambda,\lambda',h} = K(\xi_{\lambda,h}, \xi_{\lambda',h})\, w_{\lambda',h}, \tag{6.20}$$

$$\tilde{t}_{\lambda,\lambda',h} = \sum_{\{\mu\in\Lambda_{N^{-1}h}:\, G_{\mu,N^{-1}h}\subset G_{\lambda',h}\}} K(\xi_{\lambda,h}, \xi_{\mu,N^{-1}h})\, w_{\mu,N^{-1}h}, \tag{6.21}$$

omitting terms where the arguments of $K(x,y)$ are too near to one another. The integer $N\ge1$ will be chosen depending on the magnitude of $|\xi_{\lambda,h} - \xi_{\lambda',h}|$. More precisely, the following algorithms are proposed.

Algorithm 6.1. Fix numbers $c_0>0$ and $c_1>0$, find $p=p(h)\in\mathbb{N}$ such that

$$2^{-p-1}c_0 < |h| \le 2^{-p}c_0 \tag{6.22}$$

and

(i) use formula (6.20) if $|\xi_{\lambda,h}-\xi_{\lambda',h}|\geq c_0$;

(ii) use formula (6.21) with $N=2^k$ if $2^{-k}c_0\leq|\xi_{\lambda,h}-\xi_{\lambda',h}|<2^{-k+1}c_0$, $1\leq k\leq p-1$;

(iii) use formula (6.21) with $N=2^p$ if $|\xi_{\lambda,h}-\xi_{\lambda',h}|<2^{-p+1}c_0$ omitting the terms where $|\xi_{\mu,N^{-1}h}-\xi_{\lambda,h}|<c_1 2^{-p}|h|$.

Algorithm 6.2 differs from Algorithm 6.1 only in prescription (ii) which now has the form:

(ii') use formula (6.21) with $N=2^{k-\sigma_k}$ if $2^{-k}c_0\leq|\xi_{\lambda,h}-\xi_{\lambda',h}|<2^{-k+1}c_0$, $1\leq k\leq p-1$, where $\sigma_k=[s\log_2 k]$ is the integer part of $s\log_2 k$ and $s>1/n$ is a further parameter.

Due to (6.22), the smallest mesh size used in Algorithms 6.1 and 6.2 is of the order $|h|^2$.

Lemma 6.1. Assume that, for any $h\in H_C$, every non-standard weight $w_{\lambda,h}$ corresponding to $\lambda\in\Lambda_h$ with $\mathrm{dist}(\xi^0_{\lambda,h},\partial G)\leq|h|$ and satisfying (6.7) can be found in $\mathrm{const}\,|h|^{-1}$ arithmetical operations where the constant is independent of h. Then the amount of the work to evaluate l_h^2 integrals (6.19) by the means of Algorithm 6.1 or 6.2 is, respectively, $\mathcal{O}(l_h^2\log_2 l_h)$ and $\mathcal{O}(l_h^2)$ arithmetical operations.

Proof. First we estimate the amount of the work which is needed to compute the non-standard weights. For fixed h, they correspond to $G_{\lambda,h}$ in the $2|h|$-layer of ∂G. Due to (6.2), the number of the non-standard weights does not exceed $c|h|/\mathrm{meas}\,B_{\lambda,h}=\mathcal{O}(|h|^{1-n})$, and according to the condition of the Lemma they can be found in $c|h|^{-n}$ arithmetical operations. Algorithm 6.1 uses mesh sizes $2^{-k}|h|$, $k=0,1,\dots,p$. Thus, the work to find all non-standard weights used in Algorithm 6.1 is estimated by

$$c|h|^{-n}\sum_{k=0}^{p}2^{kn}\leq c'|h|^{-n}2^{pn}\leq c''|h|^{-2n}\leq c'''l_h^2$$

arithmetical operations (we exploited (6.22) here). Algorithm 6.2 needs slightly less work but still $O(l_h^2)$ arithmetical operations to evaluate the weights.

Let us consider the case of Algorithm 6.1. It is sufficient to show that, for any fixed $\lambda\in\Lambda_h$, the elements $\tilde{t}_{\lambda,\lambda',h}$ $(\lambda'\in\Lambda_h)$ can be calculated in $O(l_h\log_2 l_h)$ operations. It is clear that the calculations via (6.20) take $O(l_h)$ operations. Further, every applying of (6.21) with $N=2^k$ $(1\leq k\leq p-1)$ costs $\leq 2\cdot 2^{kn}$ operations (here 2^{kn} is the number of nodes used in (6.21)). Thereby, formula (6.21) with $N=2^k$ is used not more than

$$\mathrm{meas}\{y\in\mathbb{R}^n:\ 2^{-k}c_0-|h|\leq|\xi_{\lambda,h}-y|\leq 2^{-k+1}c_0+|h|\}/\mathrm{meas}\,B_{\lambda,h}\equiv\tau_{k,h}$$

times, and this quantity can be estimated as follows:

$$\tau_{k,h} \le c|h|^{-n}\text{meas}\{y\in\mathbb{R}^n: |y|\le 2^{-k+2}c_0\} \le c'|h|^{-n}2^{-kn}.$$

Thus, the total work with (6.21) with $N=2^k$ is $c|h|^{-n}$ arithmetical operations and, for all $k=1,\dots,p-1$, this number is $c'|h|^{-n}|\log_2|h||$ since , due to (6.22), $p \le |\log_2|h||+\log_2 c_0$. Quantities $|h|^{-n}|\log_2|h||$ and $I_h \log_2 I_h$ are of the same order. It remains to estimate the amount of the work with $N=2^p$. Due to (6.22), condition $|\xi_{\lambda,h}-\xi_{\lambda',h}| < 2^{-p+1}c_0$ implies inequality $|\xi_{\lambda,h}-\xi_{\lambda',h}| < 4|h|$, therefore the number of $\tilde{t}_{\lambda,\lambda',h}$ calculated via $\{(6.21), N=2^p\}$ remains bounded as $|h|\to 0$. One evaluation by $\{(6.21), N=2^p\}$ costs not more than $2\cdot 2^{pn} \le 2(c_0/|h|)^n \le cI_h$ arithmetical operations, thus the total amount of work is $O(I_h)$ arithmetical operations. This completes the proof of the Lemma for Algorithm 6.1.

Let us consider the case of Algorithm 2. Let $\lambda\in\Lambda_h$ be fixed again. The number of arithmetical operations on every application of (6.21) with $N=2^{k-\sigma_k}$ is $2\cdot 2^{(k-\sigma_k)n} \le 2\cdot 2^{kn}\cdot(2k^{-s})^n$; the number of evaluations was estimated by $c'|h|^{-n}2^{-kn}$. Thus, all applications of (6.21) with $N=2^{k-\sigma_k}$ for $k=1,\dots,p-1$, but fixed $\lambda\in\Lambda_h$, are done in

$$c|h|^{-n}\sum_{k=1}^{p-1}k^{-ns} \le c'|h|^{-n} \le c''I_h$$

arithmetical operations. This completes the proof of the Lemma for Algorithm 6.2.

6.4. Error analysis (preliminaries). Here we examine the preciseness of the cubature approximation for a single coefficent $t_{\lambda,\lambda',h}$.

Lemma 6.2. Let (3.2) with $0<\nu<n$, conditions (6.1) - (6.7) and (PS) be satisfied. Let $|\xi_{\lambda,h}-\xi_{\lambda',h}| \ge 2|h|$. Then, for cubature approximation (6.21),

$$|t_{\lambda,\lambda',h}-\tilde{t}_{\lambda,\lambda',h}| \le \text{const}\, N^{-2}|h|^2 \int_{G_{\lambda',h}} |\xi_{\lambda,h}-y|^{-\nu-2}\,dy$$
$$+ \text{const}\, N^{-1}|h| \int_{\{y\in B_{\lambda',h}:\rho(y)<N^{-1}|h|\}} |\xi_{\lambda,h}-y|^{-\nu-1}dy \qquad (\lambda,\lambda'\in\Lambda_h) \tag{6.23}$$

(if $G_{\lambda',h}=B_{\lambda',h}$ then the second term in the right hand side may be cancelled).

Proof. We have, due to (6.7),

$$|t_{\lambda,\lambda',h}-\tilde{t}_{\lambda,\lambda',h}|$$
$$\le \sum_{\{\mu\in\Lambda_{N^{-1}h}:\, G_{\mu,N^{-1}h}\subset G_{\lambda',h}\}} \left|\int_{G_{\mu,N^{-1}h}} \left[K(\xi_{\lambda,h},y)-K(\xi_{\lambda,h},\xi_{\mu,N^{-1}h})\right]dy\right| +$$

$$+ \sum_{\{\mu\in\Lambda_{N^{-1}h}:\, G_{\mu,N^{-1}h}\subset G_{\lambda,h},\, \mathrm{dist}(\xi^0_{\mu,N^{-1}h},\partial G)\le N^{-1}|h|\}} |K(\xi_{\lambda,h},\xi_{\mu,N^{-1}h})|$$

$$\cdot \left|w_{\mu,N^{-1}h} - \mathrm{meas}\, G_{\mu,N^{-1}h}\right|. \quad (6.24)$$

For the inner boxes $G_{\mu,N^{-1}h} = B_{\mu,N^{-1}h} \subset G$, thanks to (6.6),

$$\int_{G_{\mu,N^{-1}h}} (y - \xi_{\mu,N^{-1}h})\, dy = 0,$$

therefore, for those $G_{\mu,N^{-1}h}$,

$$\left| \int_{G_{\mu,N^{-1}h}} \left[K(\xi_{\lambda,h}, y) - K(\xi_{\lambda,h}, \xi_{\mu,N^{-1}h})\right] dy \right|$$

$$= \left| \int_{G_{\mu,N^{-1}h}} \left[K(\xi_{\lambda,h}, y) - K(\xi_{\lambda,h}, \xi_{\mu,N^{-1}h}) - \frac{\partial K(\xi_{\lambda,h}, \xi_{\mu,N^{-1}h})}{\partial y} (y - \xi_{\mu,N^{-1}h})\right] dy \right|$$

$$\le \frac{1}{2} \int_{G_{\mu,N^{-1}h}} \max_{0\le t\le 1} \left| \frac{\partial^2 K(\xi_{\lambda,h}, ty + (1-t)\xi_{\mu,N^{-1}h})}{\partial y^2} \right| |y - \xi_{\mu,N^{-1}h}|^2\, dy$$

where the derivatives are understood in the sense of Fréchet. Here $|y - \xi_{\mu,N^{-1}h}|^2 \le N^{-2}|h|^2/4$ for $y \in G_{\mu,N^{-1}h}$ and, as a consequence of (3.2),

$$\left| \frac{\partial^2 K(\xi_{\lambda,h}, ty + (1-t)\xi_{\mu,N^{-1}h})}{\partial y^2} \right| \le c\, |\xi_{\lambda,h} - (ty + (1-t)\xi_{\mu,N^{-1}h})|^{-\nu-2}$$

$$\le c' |\xi_{\lambda,h} - y|^{-\nu-2}, \qquad 0 \le t \le 1.$$

Summing up over μ we obtain the first term in the right hand side of estimate (6.23).

Now consider the boxes $B_{\mu,N^{-1}h}$ with $\mathrm{dist}(\xi^0_{\mu,N^{-1}h}, \partial G) \le N^{-1}|h|$. For these boxes we use a more simple estimate

$$\left| \int_{G_{\mu,N^{-1}h}} \left[K(\xi_{\lambda,h}, y) - K(\xi_{\lambda,h}, \xi_{\mu,N^{-1}h})\right] dy \right|$$

$$\le c\, N^{-1}|h| \int_{G_{\mu,N^{-1}h}} |\xi_{\lambda,h} - y|^{-\nu-1}\, dy$$

which is a corollary of (3.2) and (6.5) . Summing up over those μ we obtain the second term in the right hand side of estimate (6.23).

Further, due to (6.7), the second sum in (6.24) can be bounded by

$$cN^{-1}|h| \sum_{\{\mu\in\Lambda_{N^{-1}h}:\, G_{\mu,N^{-1}h}\subset G_{\lambda',h},\ \operatorname{dist}(\xi^0_{\mu,N^{-1}h},\partial G)\le N^{-1}|h|\}} \int_{B_{\mu,N^{-1}h}} |K(\xi_{\lambda,h},\ \xi_{\mu,N^{-1}h})|\,dy.$$

Estimating $|K(x,y)|$ by $c|x-y|^{-\nu-1}$ (not by $b|x-y|^{-\nu}$ that were also possible) we represent this quantity also in the form of second term in the right hand side of (6.23). Thereby we exploit the fact that the $|\xi_{\lambda,h}-\xi_{\mu,N^{-1}h}|$ and $|\xi_{\lambda,h}-y|$ are of the same order if $y\in B_{\mu,N^{-1}h}$ and $|\xi_{\lambda,h}-\xi_{\mu,N^{-1}h}|\ge 2N^{-1}|h|$. The proof of Lemma 6.2 is completed.

Now we consider the case where $\xi_{\lambda,h}$ and $\xi_{\lambda',h}$ may be close to one another, e.g. $\lambda=\lambda'$.

Lemma 6.3. Let (3.2) with $0<\nu<n$, (6.1)-(6.7) and (PS) be satisfied. Omit from (6.21) the terms with $|\xi_{\mu,N^{-1}h}-\xi_{\lambda,h}|<c_1N^{-1}|h|$. Then

$$\left|t_{\lambda,\lambda',h}-\tilde t_{\lambda,\lambda',h}\right| \le \text{const}\,(N^{-1}|h|)^{n-\nu}$$

$$+\,\text{const}\,N^{-2}|h|^2 \int_{\{y\in G_{\lambda',h}:\,|\xi_{\lambda,h}-y|>N^{-1}|h|\}} |\xi_{\lambda,h}-y|^{-\nu-2}\,dy \tag{6.25}$$

$$+\,\text{const}\,N^{-1}|h| \int_{\{y\in B_{\lambda',h}:\ \rho(y)<N^{-1}|h|,\ |\xi_{\lambda,h}-y|>N^{-1}|h|\}} |\xi_{\lambda,h}-y|^{-\nu-1}\,dy.$$

Proof. We estimate the integrals and their cubature approximations in a rough manner if the arguments of $K(x,y)$ are too near to one another:

$$\Bigl|\sum_{\{\mu\in\Lambda_{N^{-1}h}:\ G_{\mu,N^{-1}h}\subset G_{\lambda',h},\,|\xi_{\lambda,h}-\xi_{\mu,N^{-1}h}|<2N^{-1}|h|\}} \int_{G_{\mu,N^{-1}h}} K(\xi_{\lambda,h},\ y)\,dy\Bigr|$$

$$\le b\int_{\{y\in\mathbb{R}^n:\,|\xi_{\lambda,h}-y|<3N^{-1}|h|\}} |\xi_{\lambda,h}-y|^{-\nu}\,dy \le \text{const}\,(N^{-1}|h|)^{n-\nu},$$

$$\Bigl|\sum_{\{\mu\in\Lambda_{N^{-1}h}:\,G_{\mu,N^{-1}h}\subset G_{\lambda',h},\ c_1N^{-1}|h|\le|\xi_{\lambda,h}-\xi_{\mu,N^{-1}h}|<2N^{-1}|h|\}} K(\xi_{\lambda,h},\ \xi_{\mu,N^{-1}h})\,w_{\mu,N^{-1}h}\Bigr|$$

$$\le b\,(c_1N^{-1}|h|)^{-\nu}\,\text{meas}\{y\in\mathbb{R}^n:\,|\xi_{\lambda,h}-y|<3N^{-1}|h|\}\le\text{const}\,(N^{-1}|h|)^{n-\nu}$$

(the last sum occurs in (6.21) only in the case $c_1<2$). After that the remaining terms can be treated in a similar way as in the proof of Lemma 6.2, and the result is (6.25).

6.5. Error analysis of Algorithms 6.1 and 6.2. Introduce the matrices

$$T_h=(t_{\lambda,\lambda',h})_{\lambda,\lambda'\in\Lambda_h}, \qquad \tilde{T}=(\tilde{t}_{\lambda,\lambda',h})_{\lambda,\lambda'\in\Lambda_h}$$

where $\tilde{t}_{\lambda,\lambda',h}$ $(\lambda,\lambda'\in\Lambda_h)$ are computed by means of Algorithm 6.1 or 6.2. We shall estimate the norm

$$\|T_h-\tilde{T}_h\| = \max_{\lambda\in\Lambda_h}\sum_{\lambda'\in\Lambda_h}|t_{\lambda,\lambda',h}-\tilde{t}_{\lambda,\lambda',h}|\,.$$

<u>Lemma</u> 6.4. Let (3.2), (6.1)-(6.7) and (PS) be satisfied. Then, for Algorithms 6.1 and 6.2,

$$\|T_h-\tilde{T}_h\| \le \text{const}\,(|h|^2+|h|^{2(n-\nu)}). \tag{6.26}$$

<u>Proof</u>. It sufficies to prove (6.26) in the case $n-1\le\nu<n$. Indeed, if (3.2) holds for with a $\nu<n-1$ then (3.2) holds for $\nu=n-1$ also, and (6.26) with $\nu=n-1$ provides $\|T_h-\tilde{T}_h\|\le\text{const}\,|h|^2$. Thus, we assume that $n-1\le\nu<n$.

Let us prove (6.26) for Alghorithm 1. We have

$$\|T_h-\tilde{T}_h\| = \max_{\lambda\in\Lambda_h}\sum_{\lambda'\in\Lambda_h}|t_{\lambda,\lambda',h}-\tilde{t}_{\lambda,\lambda',h}| = \max_{\lambda\in\Lambda_h}\sum_{k=0}^{p}\tau^{(k)}_{\lambda,h}$$

where the terms

$$\tau^{(0)}_{\lambda,h} = \sum_{\{\lambda'\in\Lambda_h:\ |\xi_{\lambda,h}-\xi_{\lambda',h}|\ge c_0\}}|t_{\lambda,\lambda',h}-\tilde{t}_{\lambda,\lambda',h}|\,,$$

$$\tau^{(k)}_{\lambda,h} = \sum_{\{\lambda'\in\Lambda_h:\ 2^{-k}c_0\le|\xi_{\lambda,h}-\xi_{\lambda',h}|<2^{-k+1}c_0\}}|t_{\lambda,\lambda',h}-\tilde{t}_{\lambda,\lambda',h}|\,,$$

$$k=1,\dots,p-1,$$

$$\tau^{(p)}_{\lambda,h} = \sum_{\{\lambda'\in\Lambda\ :\ |\xi_{\lambda,h}-\xi_{\lambda',h}|<2^{-p+1}c_0\}}|t_{\lambda,\lambda',h}-\tilde{t}_{\lambda,\lambda',h}|$$

correspond to different definitions of $\tilde{t}_{\lambda,\lambda',h}$, see (i), (ii) and (iii). Denote by d the diameter of G. Using Lemmas 6.2 and 6.3 we estimate

$$\tau^{(0)}_{\lambda,h}\le c|h|^2\int_{\{y\in\mathbb{R}^n:\ c_0-|h|<|\xi_{\lambda,h}-y|<d\}}|\xi_{\lambda,h}-y|^{-\nu-2}\,dy+$$

$$+ c|h| \int_{\{y\in\mathbb{R}^n:\ c_0-|h|<|\xi_{\lambda,h}-y|<d,\ \rho(y)<|h|\}} |\xi_{\lambda,h}-y|^{-\nu-1} \le c'|h|^2,$$

$$\tau^{(k)}_{\lambda,h} \le c\,2^{-2k}|h|^2 \int_{\{y\in\mathbb{R}^n:\ 2^{-k-1}c_0<|\xi_{\lambda,h}-y|<2^{-k+2}c_0\}} |\xi_{\lambda,h}-y|^{-\nu-2}\,dy$$

$$+ c\,2^{-k}|h| \int_{\{y\in\mathbb{R}^n:\ 2^{-k-1}c_0<|\xi_{\lambda,h}-y|<2^{-k+2}c_0,\ \rho(y)<2^{-k}|h|\}} |\xi_{\lambda,h}-y|^{-\nu-1}dy$$

$$\le c'2^{-k(n-\nu)}|h|^2 \qquad (k=1,\ \dots\ ,p-1), \tag{6.27}$$

$$\tau^{(p)}_{\lambda,h} \le c(2^{-p}|h|)^{n-\nu}$$

$$+ c\,2^{-2p}|h|^2 \int_{\{y\in\mathbb{R}^n:\ 2^{-p}|h|<|\xi_{\lambda,h}-y|<2^{-p+2}c_0\}} |\xi_{\lambda,h}-y|^{-\nu-2}$$

$$+ c\,2^{-p}|h| \int_{\{y\in\mathbb{R}^n:\ 2^{-p}|h|<|\xi_{\lambda,h}-y|<2^{-p+2}c_0,\ \rho(y)<2^{-p}|h|\}} |\xi_{\lambda,h}-y|^{-\nu-1}dy$$

$$\le c'|h|^{2(n-\nu)}.$$

Here we took into account that, due to (6.22), $2^{-p} \le 2c_0^{-1}|h|$, and for $\nu \ge n-1$,

$$\int_{\{y\in\mathbb{R}^n:\ r<|y|<d\}} |y|^{-\nu-2}dy \le \text{const}\ r^{n-\nu-2},$$

$$\int_{\{y'\in\mathbb{R}^{n-1}:\ r<|y'|<d\}} |y'|^{-\nu-1}dy' \le \text{const}\ r^{n-\nu-2};$$

we also have

$$\int_{\{y\in\mathbb{R}^n:\ r<|\xi_{\lambda,h}-y|<r_1,\ \rho(y)<\delta\}} |\xi_{\lambda,h}-y|^{-\nu-1}dy$$

$$\le c\delta \int_{\{y'\in\mathbb{R}^{n-1}:\ r<|y'|<r_1\}} |y'|^{-\nu-1}dy'$$

(this inequality can be established by arguments of the rectification of the boundary ∂G as in Section 5.1).

Summing up we obtain

$$\left\| T_h - \tilde{T}_h \right\| \le c'(|h|^2 + \sum_{k=1}^{p-1} 2^{-k(n-\nu)}|h|^2 + |h|^{2(n-\nu)}) \le c|h|^2 + c'|h|^{2(n-\nu)}.$$

This complectes the proof in the case of Algorithm 6.1.

In the case of Algorithm 6.2, only inequalities (6.27) must be overlooked. Now, instead of 2^{-2k} and 2^{-k}, multipliers $2^{-2(k-\sigma_k)}$ and $2^{-(k-\sigma_k)}$ arise in front of integrals in (6.27), and the result is

$$\tau_{\lambda,h}^{(k)} \le c' k^{2s} 2^{-k(n-\nu)} |h|^2 \qquad (k=1,\dots,p-1).$$

Consequently, (6.26) holds again. The proof of Lemma 6.4 is completed.

6.6. Error estimates for approximate solution. We are interested in the behavior of the solution of system (6.9) with approximated coefficients:

$$\tilde{u}_{\lambda,h} = \sum_{\lambda' \in \Lambda_h} \tilde{t}_{\lambda,\lambda',h}\, \tilde{u}_{\lambda',h} + f(\xi_{\lambda,h}), \qquad \lambda \in \Lambda_h. \tag{6.28}$$

Solving this system we can define an approximate solution to (3.1) for all $x \in G$ in a similar way as in (6.13):

$$\tilde{u}_h(x) = \sum_{\lambda' \in \Lambda_h} \int_{G_{\lambda',h}} K(x,y)dy\, \tilde{u}_{\lambda',h} + f(x), \qquad x \in G. \tag{6.29}$$

Integrals in (6.29) can be calculated using an extension of Algorithm 6.1 or 6.2 which we obtain subtituting $\xi_{\lambda,h}$ for x in (6.20), (6.21) and (i)–(iii). Thus we design an approximation

$$\tilde{v}_h(x) = \sum_{\lambda' \in \Lambda_h} \tilde{t}_{\lambda',h}(x)\, \tilde{u}_{\lambda',h} + f(x), \qquad x \in G. \tag{6.30}$$

Every evaluation of $\tilde{v}_h(x)$ at a point $x \in G$, $x \notin \Xi_h$, costs $O(l_h)$ or $O(l_h \log_2 l_h)$ arithmetical operations if Algorithm 6.1, respectively, Algorithm 6.2 is used. For $x = \xi_{\lambda,h}$, we have $\tilde{v}_h(\xi_{\lambda,h}) = \tilde{u}_{\lambda,h}$, $\lambda \in \Lambda_h$.

Lemma 6.5. Assume that conditions 1-3 and 5 of Theorem 6.1 are fulfilled. Let the coefficents $\tilde{t}_{\lambda,\lambda',h}$ be calculated by means of Algorithm 6.1 or Algorithm 6.2. Then there exists a $\delta_0 > 0$ such that, for all $h \in H_c$ with $|h| < \delta_0$, systems (6.9) and (6.28) are uniquely solvable, and

$$\max_{\lambda \in \Lambda_h} |u_{\lambda,h} - \tilde{u}_{\lambda,h}| \le \text{const}\,(|h|^2 + |h|^{2(n-\nu)}), \tag{6.31}$$

$$\sup_{x \in G} |u_h(x) - \tilde{u}_h(x)| \le \text{const}\,(|h|^2 + |h|^{2(n-\nu)}), \tag{6.32}$$

$$\sup_{x\in G} |\tilde{u}_h(x) - \tilde{v}_h(x)| \le \text{const } (|h|^2 + |h|^{2(n-\nu)}) \tag{6.33}$$

where $\{u_{\lambda,h}\}$ and $\{\tilde{u}_{\lambda,h}\}$ are solutions to systems (6.9) and (6.28), respectively, $u_h(x)$ is defined in (6.13), $\tilde{u}_h(x)$ is defined in (6.29) and $\tilde{v}_h(x)$ is obtained from $\tilde{u}_h(x)$ approximating the integrals in (6.29) by the means of the extension of Algorithm 6.1 or 6.2.

Proof. Introduce the space E_h of grid functions u_h: $\Xi \to \mathbb{R}$ where $\Xi_h = \{\xi_{\lambda,h}\}_{\lambda\in\Lambda}$, and equip it with the norm

$$\| u_h \| = \max_{\lambda\in\Lambda_h} |u_h(\xi_{\lambda,h})|$$

(cf. Section 4.2). Systems (6.9) and (6.28) can be represented as equations in E_h, $u_h = T_h u_h + p_h f$ and $\tilde{u}_h = \tilde{T}_h \tilde{u}_h + p_h f$, respectively, where $p_h f$ is the restriction of f to the grid Ξ_h. In Section 5.6 we proved that, under conditions 1-3 of Theorem 6.1, $T_h \dashrightarrow T$ compactly; together with condition 5 this implies (see Lemma 4.2) that the operators $I_h - T_h$ are invertible for sufficently small $|h|$ and the inverse operators are uniformly bounded in h :

$$\| (I_h - T_h)^{-1} \|_{L(E_h,E_h)} \le \text{const} \qquad (|h| < \delta_0).$$

According to Lemma 6.4, we have

$$\| T_h - \tilde{T}_h \|_{L(E_h,E_h)} \le \text{const } (|h|^2 + |h|^{2(n-\nu)})$$

(note that we estimated namely this operator norm).

The last two inequalities immediately imply (6.31). Estimation (6.32) is a direct consequence of (6.31). Repeating the arguments of the proof of Lemma 6.4 we see that

$$\sup_{x\in G} \sum_{\lambda'\in\Lambda_h} \Big| \int_{G_{\lambda',h}} K(x,y)\,dy - \tilde{t}_{\lambda',h}(x) \Big| \le \text{const}\,(|h|^2 + |h|^{2(n-\nu)}).$$

Together with the uniform boundeness of $\{\tilde{u}_{\lambda,h}\}$ as $|h| \to 0$, this implies (6.33). The proof of Lemma 6.5 is completed.

As a corollary of Theorem 6.1(a) and Lemma 6.5 we obtain the following result.

Theorem 6.2. Assume that conditions 1-5 of Theorem 6.1 are fulfilled. Then there exists a $\delta_0 > 0$ such that, for all $h \in H_C$ with $|h| < \delta_0$, the system (6.28) with the coefficients $\tilde{t}_{\lambda,\lambda',h}$ evaluated by the means of Algorithm 6.1 or 6.2 has a unique solution $\{\tilde{u}_{\lambda,h}\}_{\lambda=\Lambda_h}$, and the following error estimates hold:

$$\max_{\lambda\in\Lambda_h} |\tilde{u}_{\lambda,h} - u(\xi_{\lambda,h})| \le \text{const } (\varepsilon_{\nu,h})^2, \tag{6.34}$$

$$\sup_{x\in G} |\tilde{u}_h(x) - u(x)| \le \text{const } (\varepsilon_{\nu,h})^2 , \tag{6.35}$$

$$\sup_{x\in G} |\tilde{v}_h(x) - u(x)| \le \text{const } (\varepsilon_{\nu,h})^2 . \tag{6.36}$$

Here u is the (unique) solution of integral equation (3.1); the prolonged approximations $\tilde{u}_h$ and $\tilde{v}_h$ are defined by (6.29) and (6.30).

In other words, Algorithms 1 and 2 are sufficiently precise to preserve the convergence rate of basic method (6.9). Recall (see Lemma 6.1 for the conditions) that the Algorithm 6.2 also satisfies the requirement about the arithmetical work posed by us in the problem setting. Algorithm 6.1, being simpler, takes slightly more arithmetical work.

6.7. Some further algorithms. Cubature formulae Algorithms 6.1 and 6.2 are universal in the sense that they do not depend on ν , the strenghtness on the singularity of the kernel. In some sense, they are most properly adapted to the case $\nu = n-1$. Here we give some further modifications of Algorithm 6.1 depending on ν. In Algoritms 6.3 and 6.4, prescription (ii) allows the use of the cubature formula with an essentially smaller N than $N = 2^k$ in the case of Algorithm 6.1. On the other hand, we have not so much succeeded in preseription (iii).

Algorithm 6.3 (for $n-1 \le \nu < n$). Fix $a \in (1-(n-\nu)/2,\ 1]$, $c_0>0$, $c_1>0$, $c_2>0$, find p such that

$$2^{-p-1}c_0 < |h|^{1-\Theta} \le 2^{-p}c_0 \quad \text{with} \quad \Theta = \frac{2-2(n-\nu)}{2-(n-\nu)}$$

and

(i) use (6.20) if $|\xi_{\lambda,h} - \xi_{\lambda',h}| \ge c_0|h|^{\Theta}$;

(ii) use (6.21) with $N=2^{[ak]}$ if $2^{-k}c_0h^{\Theta} \le |\xi_{\lambda,h} - \xi_{\lambda',h}| < 2^{-k+1}c_0h^{\Theta}$, $1 \le k \le p-1$;

(iii) use (6.21) with $N=[c_1|h|^{-1}]$ if $|\xi_{\lambda,h} - \xi_{\lambda',h}| < 2^{-p+1}c_0h^{\Theta}$ omitting the terms where $|\xi_{\mu,N^{-1}h} - \xi_{\lambda,h}| < c_2|h|^2$.

Algorithm 6.4 (for $n-2 < \nu \le n-1$). Fix $a \in (1-(n-\nu)/2,\ 1]$, $c_0>0$, $c_1>0$, $c_2>0$, find p such that

$$2^{-p-1}c_0 < |h| \le 2^{-p}c_0$$

and

(i) use (6.20) if $|\xi_{\lambda,h} - \xi_{\lambda',h}| \ge c_0$;

(ii) use (6.21) with $N=2^{[ak]}$ if $2^{-k}c_0 \le |\xi_{\lambda,h} - \xi_{\lambda',h}| < 2^{-k+1}c_0$, $1 \le k \le p-1$;

(iii) use (6.21) with $N=[c_1|h|^{1-2/(n-\nu)}]$ if $|\xi_{\lambda,h}-\xi_{\lambda',h}|<2^{-p+1}c_0$ omitting the terms where $|\xi_{\mu,N^{-1}h}-\xi_{\lambda,h}|<c_2|h|^{2/(n-\nu)}$.

Algorithm 6.5 (for $\nu = n-2$). Fix $c_0>0$, $c_1>0$, find p such that

$$2^{-p-1}c_0 < |h| < 2^{-p}c_0$$

and

(i) use (6.20) if $|\xi_{\lambda,h}-\xi_{\lambda',h}| \ge c_0$;

(ii) use (6.21) with $N=k$ if $2^{-k}c_0 \le |\xi_{\lambda,h}-\xi_{\lambda',h}| < 2^{-k+1}c_0$, $1 \le k \le p-1$;

(iii) use (6.21) with $N=p$ if $|\xi_{\lambda,h}-\xi_{\lambda',h}| < 2^{-p+1}c_0$ omitting the terms where $|\xi_{\mu,N^{-1}h}-\xi_{\lambda,h}| < c_1|h|\,|\log|h||^{-1/2}$ or, in case $n \ge 3$, simply put $\tilde{t}_{\lambda,\lambda',h}=0$.

Recall that in the case $\nu < n-2$ the simplest cubature formula method (6.10) is of the accuracy $O(|h|^2)$, and no further algorithms are needed.

Lemma 6.6. Assume that the condition of Lemma 6.1 is fulfilled. Then the number of arithmetical operations to evaluate l_h^2 integrals (6.19) is as follows:

$O(l_h^2)$ for Algorithm 6.3 if $\nu > n-1$ or if $a<1$, and $O(l_h^2\log_2 l_h)$ if $\nu = n-1$, $a=1$;

$O(l_h^2)$ for Algorithm 6.4 if $a<1$, and $O(l_h^2\log_2 l_h)$ if $a=1$;

$O(l_h^2)$ for Algorithm 6.5.

Lemma 6.7. Assume that the conditions of Lemma 6.4 are fulfilled. Then:

$\|T_h-\tilde{T}_h\| \le \text{const}\ |h|^{2(n-\nu)}$ if $n-1 \le \nu < n$ and Algorithm 6.3 is applied;

$\|T_h-\tilde{T}_h\| \le \text{const}\ |h|^{2}$ if $n-2<\nu\le n-1$ and Algorithm 6.4 is applied;

$\|T_h-\tilde{T}_h\| \le \text{const}\ |h|^{2}$ if $\nu=n-2$ and Algorithm 6.5 is applied.

The proof of these assertions are similar to the proofs of Lemmas 6.1 and 6.4.

From Lemma 6.7 it follows again that Algorithms 6.3, 6.4 and 6.5 preserve the convergence rate of the basic method (6.9) for respective ν.

6.8. Two grid method. Let $h_* = Nh$ where $N \gg 1$ is an integer. Note that every "panel" G_{λ',h_*} is a union of cells $G_{\lambda,h}$ such that $\xi_{\lambda,h}\in G_{\lambda',h_*}$. For

$$\Xi_h = \{\xi_{\lambda,h}\}_{\lambda\in\Lambda_h}, \qquad \Xi_{h_*} = \{\xi_{\lambda',h_*}\}_{\lambda'\in\Lambda_{h_*}}$$

we define the operators

$$\Pi_{h_*h}:\Xi_h \to \Xi_{h_*}, \qquad \Pi_{hh_*}:\Xi_{h_*} \to \Xi_h$$

requiring that the following holds:

for $\xi_{\lambda,h} \in \Xi_h$, $\Pi_{h_*h}\xi_{\lambda,h} = \xi_{\lambda',h_*}$ is such that $G_{\lambda,h} \subset G_{\lambda',h_*}$;

for $\xi_{\lambda',h_*} \in \Xi_{h_*}$, $\Pi_{hh_*}\xi_{\lambda',h_*} = \xi_{\lambda,h}$ is such that $G_{\lambda,h} \subset G_{\lambda',h_*}$.

The operator Π_{h_*h} is uniquely defined by the first requirement but not Π_{hh_*} by the second one; we somehow fix a possible Π_{hh_*} satisfying the second requirement. For instance, if $\Xi_{h_*} \subset \Xi_h$ (as in Section 5) then the definition $\Pi_{hh_*}\xi_{\lambda',h_*} = \xi_{\lambda',h_*}$, $\lambda' \in \Lambda_{h_*}$, is possible. Now define the connection operators

$$p_{h_*h} \in \mathcal{L}(E_h, E_{h_*}), \qquad p_{hh_*} \in \mathcal{L}(E_{h_*}, E_h)$$

as follows:

$$(p_{h_*h}u_h)(\xi_{\lambda',h_*}) = u_h(\Pi_{hh_*}\xi_{\lambda',h_*}), \qquad \lambda' \in \Lambda_{h_*},\ u_h \in E_h,$$

$$(p_{hh_*}u_{h_*})(\xi_{\lambda,h}) = u_{h_*}(\Pi_{h_*h}\xi_{\lambda,h}), \qquad \lambda \in \Lambda_h,\ u_{h_*} \in E_{h_*}.$$

To solve the system (6.28) we introduce the two grid method (cf. (5.41))

$$\begin{aligned} \tilde{v}_h^k &= \tilde{u}_h^k - \tilde{T}_h\tilde{u}_h^k - p_h f, \\ \tilde{u}_h^{k+1} &= \tilde{u}_h^k - \tilde{v}_h^k - p_{hh_*}(I_{h_*} - \tilde{T}_{h_*})^{-1} p_{h_*h}\tilde{T}_h\tilde{v}_h^k, \quad k = 0,1,2,\dots . \end{aligned} \tag{6.37}$$

An equivalent form of iterations (6.37) is

$$\tilde{u}_h^{k+1} = \tilde{T}_{h,h_*}\tilde{u}_h^k + \tilde{f}_{h,h_*}, \quad k = 0,1,2,\dots$$

where

$$\tilde{T}_{h,h_*} = (I_h - p_{hh_*}p_{h_*h})\tilde{T}_h + p_{hh_*}(I_{h_*} - \tilde{T}_{h_*})^{-1}(p_{h_*h}\tilde{T}_h - \tilde{T}_{h_*}p_{h_*h})\tilde{T}_h \in \mathcal{L}(E_h, E_h),$$

$$\tilde{f}_{h,h_*} = p_h f + p_{hh_*}(I_{h_*} - \tilde{T}_{h_*})^{-1} p_{h_*h}\tilde{T}_h p_h f \in E_h.$$

<u>Theorem</u> 6.3. Assume that conditions 1-5 of Theorem 6.1 are fulfilled. Let the elements $\tilde{t}_{\lambda,\lambda',h}$ $(\lambda,\lambda' \in \Lambda_h)$ of $\tilde{T}_h$ and the elements $\tilde{t}_{\lambda,\lambda',h_*}$ $(\lambda,\lambda' \in \Lambda_{h_*})$ of $\tilde{T}_{h_*}$ be evaluated by Algorithm 6.1 or 6.2 (or, for corresponding ν, by Algorithms 6.3, 6.4 or 6.5). Then there is a $\delta_0 > 0$ such that, for $h \in H_C$, $0 < |h| < \delta_0$, we have $\|\tilde{T}_{h,h_*}\| \le c\varepsilon_{\nu,h_*} < 1$ and

$$\|\tilde{u}_h^k - \tilde{u}_h\| \le \|\tilde{u}_h^0 - \tilde{u}_h\|(c\varepsilon_{\nu,h_*})^k, \quad k = 1,2,\dots$$

where $\tilde{u}_h$ is the exact solution of system (6.28), $\tilde{u}_h^k$ are defined by iteration method (6.37) and the constant c is independent of h and h_*.

<u>Proof</u>. Repeating the proof of Theorem 5.2 we first establish a similar

result for the iterations with exact T_h, T_{h_*} and the solution of (6.9). After that we use Lemmas 5.7 and 6.4 (or 6.7). A detailed argument is proposed to the reader.

The analysis of the amount of the arithmetical work to solve (6.28) with an accuracy $\mathcal{O}(\varepsilon^2_{\nu,h})$ fully repeats the arguments of Section 5.14 – $\mathcal{O}(l_h^2)$ arithmetical operations are sufficient again. Thus, the full cycle of the numerical implementation of the PCCM - the cubature evaluation of the coefficients of the system as well the solution of the system - can be done in $\mathcal{O}(l_h^2)$ arithmetical operations maitaining the accuracy $\mathcal{O}(\varepsilon^2_{\nu,h})$ of the basic method.

6.9. Case of convolution type integral equations. Often the kernel $K(x,y)$ of equation (3.1) has the form

$$K(x,y) = a(x)\varkappa(x-y)b(y) \tag{6.38}$$

where $a,b \in BC^2(G)$ and $\varkappa(x)$ has a singularity at $x=0$. Condition (3.2) takes the form $(m=2)$

$$|D^\alpha \varkappa(x)| \le b \begin{Bmatrix} 1 & , & \nu+|\alpha|<0 \\ 1+|\log|x||, & & \nu+|\alpha|=0 \\ |x|^{-\nu-|\alpha|} & , & \nu+|\alpha|>0 \end{Bmatrix}, \quad |\alpha| \le 2. \tag{6.39}$$

The approximation

$$t'_{\lambda,\lambda',h} = a(\xi_{\lambda,h}) \int_{G_{\lambda',h}} \varkappa(\xi_{\lambda,h}-y)\,dy\, b(\xi_{\lambda',h}), \quad \lambda,\lambda' \in \Lambda_h,$$

is of the accuracy (see Lemma 5.4)

$$\|T'_h - T_h\| \le \text{const}\, \varepsilon^2_{\nu,h}.$$

Approximating the integrals

$$\int_{G_{\lambda',h}} \varkappa(\xi_{\lambda,h}-y)\,dy, \quad \lambda,\lambda' \in \Lambda_h, \tag{6.40}$$

by Algorithms 6.1 or 6.2 (or by 6.3–6.5 for corresponding ν) we obtain an approximation $\tilde{T}'_h$ of the accuracy (see Lemmas 6.4 and 6.7)

$$\|\tilde{T}'_h - T_h\| \le \text{const}\, \varepsilon^2_{\nu,h}.$$

Consider the case where G is a parallelepided:

$$G = \{x \in \mathbb{R}^n:\ 0<x_i<b_i,\ i=1,\dots,n\}.$$

Then we can use regular partitions of G with $h_i = b_i/N_i$, $N_i \in \mathbb{N}$ $(i=1,\dots,n)$, up to the boundary:

$$G_{\lambda,h} = B_{\lambda,h}, \quad \xi_{\lambda,h} = \xi^{0}_{\lambda,h}, \quad \lambda \in \Lambda_h = \{\lambda \in \mathbb{Z}^n : \xi^{0}_{\lambda,h} \in G\}$$

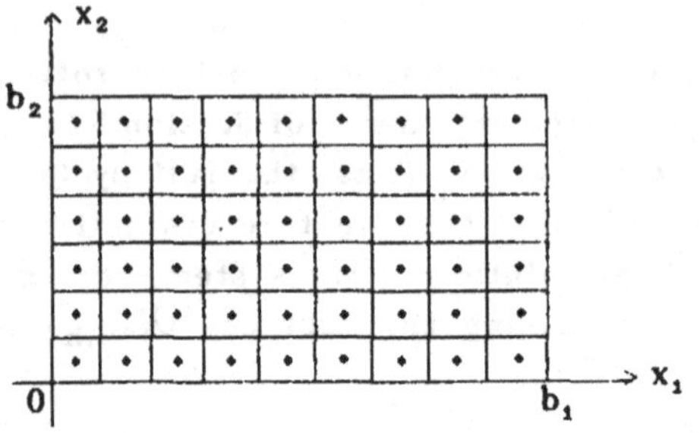

Fig. 6.3

(see Figure 6.3). The integrals (6.40) take the form

$$\int_{G_{\lambda',h}} \varkappa(\xi_{\lambda,h}-y)\,dy = \int_{B_{\lambda',h}} \varkappa(\xi^{0}_{\lambda,h}-y)\,dy = \int_{B_{\lambda-\lambda',h}} \varkappa(z)\,dz = s_{\lambda-\lambda',h}, \quad \lambda,\lambda' \in \Lambda_h.$$

The evaluation of the coefficients $s_{\lambda-\lambda',h}$, $\lambda,\lambda' \in \Lambda_h$, by Algorithm 6.1 or 6.2 takes, respectively, $\mathcal{O}(l_h \log_2 l_h)$ and $\mathcal{O}(l_h)$ arithmetical operations; a similar result holds for Algorithms 6.3-6.5. Further, the system $\tilde{u}_h = \tilde{T}'_h \tilde{u}_h + p_h f$ is of the convolution type:

$$\tilde{u}_{\lambda,h} = \sum_{\lambda' \in \Lambda_h} a(\xi_{\lambda,h})\tilde{s}_{\lambda-\lambda',h} b(\xi_{\lambda',h})\tilde{u}_{\lambda',h} + f(\xi_{\lambda,h}), \quad \lambda \in \Lambda_h, \tag{6.41}$$

and an application of $\tilde{T}'_h = (p_h a)\tilde{S}_h(p_h b)$, more precisely, of $\tilde{S}_h$ to a $v_h \in E_h$ can be realized via FFT in $\mathcal{O}(l_h \log_2 l_h)$ arithmetical operations. System (6.41) can be solved with $\mathcal{O}(\varepsilon^2_{\nu,h})$ accuracy in $\mathcal{O}(l_h \log_2 l_h)$ arithmetical operations starting from the initial guess $\tilde{u}^0_h = p_{hh_*}(I_{h_*} - \tilde{T}'_{h_*})^{-1} p_{h_*} f$ and performing six steps of the two grid iteration method which is similar to (6.37), with $\tilde{T}'_h$ instead of $\tilde{T}_h$ and with $|h_*| \sim |h|^{1/3}$, i.e. $h_* = Nh$, $\mathbb{N} \ni N \sim |h|^{-2/3}$ (cf. with Section 5.14). Thus, in the case of the integral equation (3.1) on the parallelepiped, with the kernel of the form (6.38), (6.39)), a full numerical implementation of the basic PCCM maintaining its accuracy $\mathcal{O}(\varepsilon^2_{\nu,h})$ can be done in $\mathcal{O}(l_h \log_2 l_h)$ arithmetical operations.

6.10. Exercises. 6.10.1. In Algorithm 6.1 and other algorithms, more levels of $|\xi_{\lambda,h} - \xi_{\lambda',h}|$ may be introduced. Prove that the results of the Chapter remain true if the prescription (ii) of Algorithm 1 is replaced by the following one: use formula (6.21) with $N = k+1$ if

$$c_0/(k+1) \le |\xi_{\lambda,h} - \xi_{\lambda',h}| < c_0/k, \quad 1 \le k \le 2^{p-1}-1.$$

6.10.2. Prove Lemma 6.6.

6.10.3. Prove Lemma 6.7.

6.10.4. Present a detailed proof of Theorem 6.3.

6.10.5. Prove that Theorem 6.3 remains true evaluating $\tilde{t}_{\lambda,\lambda',h}$ $(\lambda,\lambda' \in \Lambda_h)$ by Algorithm 6.1 or 6.2 but defining

$$\tilde{t}_{\lambda,\lambda',h_*} = \sum_{\{\lambda'' \in \Lambda_h:\, G_{\lambda'',h} \subset G_{\lambda',h_*}\}} \tilde{t}_{\mu,\lambda'',h}, \qquad \lambda,\lambda' \in \Lambda_{h_*},$$

where $\mu = \mu(\lambda) \in \Lambda_h$, $\lambda \in \Lambda_{h_*}$, is such that $\Pi_{hh_*}\xi_{\lambda,h_*} = \xi_{\mu,h}$.

7. HIGHER ORDER METHODS

In this Chapter, for integral equation (3.1) on a parallelepiped, collocation methods of an arbitrary order $\mathcal{O}(h^m)$ are constructed. Graded grids and piecewise polynomial approximations are used. The convergence analysis is based on Theorem 3.2.

7.1. Space $C_{\square}^{m,\nu}(G)$. Let $G \subset \mathbb{R}^n$ be a parallelepiped,

$$G = \{x \in \mathbb{R}^n : 0 < x_k < b_k,\ k = 1, \dots, n\}. \tag{7.1}$$

Denoting, for $x \in G$,

$$\rho_k(x) = \min\{x_k, b_k - x_k\}, \quad k = 1, \dots, n,$$

we have

$$\rho(x) = \operatorname{dist}(x, \partial G) = \min_{1 \le k \le n} \rho_k(x).$$

For a $\nu \in \mathbb{R}$, $\nu < n$, introduce the space $C_{\square}^{m,\nu}(G)$ consisting of functions $u \in C^{m,\nu}(G)$ (see Section 2.5) such that

$$\left|\frac{\partial^l u(x)}{\partial x_k^l}\right| \le \text{const} \begin{cases} 1 & , \ l < n-\nu \\ 1 + |\log \rho_k(x)|, & l = n-\nu \\ \rho_k(x)^{n-\nu-l} & , \ l > n-\nu \end{cases}, \quad x \in G,\ l=1,\dots,m,\ k=1,\dots,n. \tag{7.2}$$

Note that $C^m(\overline{G}) \subset C_{\square}^{m,\nu}(G)$. Introducing the constant vector fields

$$a^k(x) = e^k = (\underbrace{0,\dots,0}_{k-1},1,\underbrace{0,\dots,0}_{n-k}), \qquad x \in \overline{G},$$

we can see that, in the designations of Section 2.6,

$$C_{\square}^{m,\nu}(G) = \bigcap_{k=1}^{n} C_{a^k}^{m,\nu}(G);$$

the vector field a^k is tangential to all surfaces of ∂G exept two ones which are perpendicular to e^k, thereby $\rho_k(x) = \rho_{a^k}(x)$. Further, due to Lemma 2.4, a function $u \in C^{m,\nu}(G)$ can be extended up to a continuous function on $\overline{G} = G^*$

(for a parallelepiped, the d_G - distance and the Euclidean distance coincide); we always assume that this extension is done.

From Theorem 3.2 we obtain the following result.

<u>Lemma</u> 7.1. If $f \in C_{\square}^{m,\nu}(G)$ and the kernel $K(x,y)$ satisfies condition (3.2) then any solution $u \in L(G)$ of integral equation (3.1) belongs to $C_{\square}^{m,\nu}(G)$.

7.2. Piecewise polynomial interpolation. In the interval $[0, b_k]$, $1 \le k \le n$, introduce the following $2N_k + 1$ grid points:

$$x_k^j = \frac{b_k}{2}\left(\frac{j}{N_k}\right)^r, \quad j = 0, 1, \dots, N_k, \qquad x_k^{N_k+j} = b_k - x_k^{N_k-j}, \quad j = 1, \dots, N_k. \tag{7.3}$$

Here $r \in \mathbb{R}$, $r \ge 1$, characterizes the degree of the non-uniformity of the grid. If $r = 1$ then the grid points are uniformly located; if $r > 1$ then the grid points are more densely located towards the end points of the interval (see Figure 7.1 where $N_k = 4$, $r = 2$). Note that $x_k^0 = 0$, $x_k^{2N_k} = b_k$ and the grid points are located symmetrically with respect to $x^{N_k} = b_k/2$. It is clear that

$$x_k^{j+1} - x_k^j \le \frac{b_k}{2} \frac{r}{N_k} \left(\frac{j+1}{N_k}\right)^{r-1}, \qquad j = 0, 1, \dots, N_k - 1, \tag{7.4}$$

and similar inequality holds for the grid points on the other half of $[0, b_k]$.

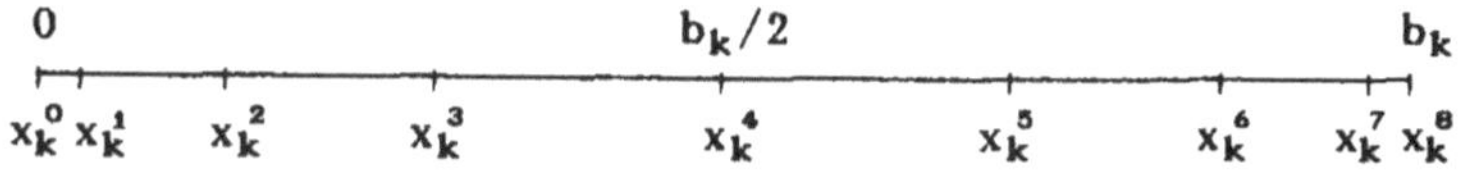

Fig. 7.1

Using points (7.3) we introduce the partition of G into the closed cells

$$G_{j_1 \dots j_n} = \{x \in \mathbb{R}^n : x_k^{j_k} \le x_k \le x_k^{j_k+1},\ k = 1, \dots, n\} \subset \overline{G},$$

$$j_k = 0, 1, \dots, 2N_k - 1, \quad k = 1, \dots, n.$$

The partition is illustrated on Figure 7.2 where $n = 2$, $N_1 = 4$, $N_2 = 3$, $r = 2$.

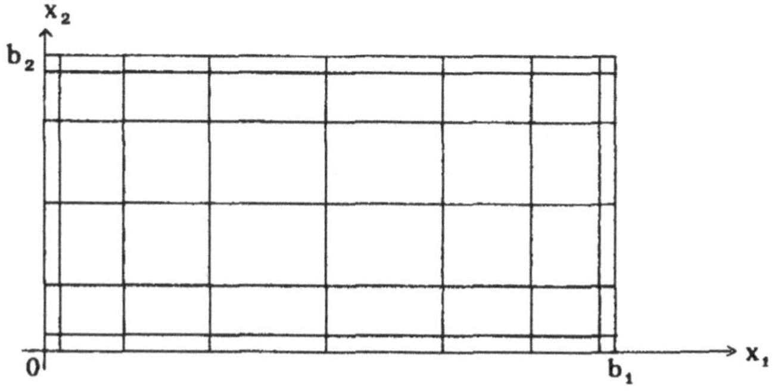

Fig. 7.2.

Let us choose in the standard interval $[-1,1]$ some interpolation points $\xi^1,\dots,\xi^m$:

$$-1 \le \xi^1 < \xi^2 < \dots < \xi^m \le 1. \tag{7.5}$$

Using affine transformations we transfer them into $[x_k^j, x_k^{j+1}]$:

$$\xi_k^{j,l} = x_k^j + \frac{\xi^l+1}{2}(x_k^{j+1} - x_k^j), \quad l=1,\dots,m;\ j=0,1,\dots,2N_k-1;\ k=1,\dots,n. \tag{7.6}$$

Note that $\xi_k^{j,m} = \xi_k^{j+1,1} = x_k^{j+1}$ if $\xi^1=-1$, $\xi^m=1$, $0 \le j \le 2N_k-2$. We assign the collocation points

$$(\xi_1^{j_1,l_1}, \dots, \xi_n^{j_n,l_n}) \in G_{j_1\dots j_n}, \quad l_k=1,\dots,m,\ k=1,\dots,n, \tag{7.7}$$

to the cell $G_{j_1\dots j_n}$. Figure 7.3 illustrates for $n=2$, $m=3$ the location of those points in $G_{j_1 j_2}$ depending on the values ξ^1 and ξ^m.

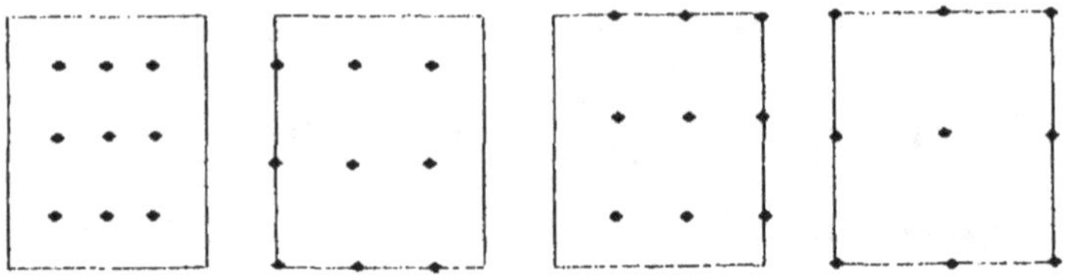

$\xi^1>-1,\ \xi^3<1$ $\quad$ $\xi^1=-1,\ \xi^3<1$ $\quad$ $\xi^1>-1,\ \xi^3=1$ $\quad$ $\xi^1=-1,\ \xi^3=1$ $\quad$ Fig.7.3

To a function $u:\overline{G}\to\mathbb{R}$ we assign a piecewise polynomial interpolation function $P_N u:\overline{G}\to\mathbb{R}$ with $N=(N_1,\dots,N_n)$ as follows: on every cell $G_{j_1\dots j_n}$ $(0\le j_k \le 2N_k-1,\ k=1,\dots,n)$, $P_N u$ is a polynomial of degree not exceeding $m-1$ with respect to any of arguments $x_1,\dots,x_n$ whereby $P_N u$ interpolates u at points (7.7). Thus, the interpolant $P_N u$ is uniquely defined in every cell separately and may have jumps on plains $x_k=x_k^j$. We may treat $P_N u$ as a multivalued function on these plains; in the sequel, the uniform error estimates concern all possible values of $P_N u$. Note that in the case $\xi^1=-1$, $\xi^m=1$ the polynomial pieces of $P_N u$ are pasted into a continuous function on $\overline{G}$, i.e. $P_N u \in C(\overline{G})$.

7.3. Error estimates of the interpolation. In addition to $N=(N_1,\dots,N_n)$, the interpolation function $P_N u$ depends on r,m and $\xi_1,\dots,\xi_m$ that is not exposed in the designations. An estimate $\|u-P_N u\| \le ch^m$ is the best that holds for a smooth function u without singularities. For $u \in C_\square^{m,\nu}(G)$, the singularities of its derivatives can be compensated using larger values of r, the scaling parameter of the grid (see (7.3)). More precisely, the following result is true.

Lemma 7.2. Assume that $u \in C_{\Box}^{m,\nu}(G)$. Then the following estimates with $h = \max_{1 \le k \le n} b_k / N_k$ hold:

if $m < n-\nu$ then
$$\max_{x \in \overline{G}} |u(x) - (P_N u)(x)| \le c h^m, \quad \forall r \ge 1; \tag{7.8}$$

if $m = n-\nu$ then
$$\max_{x \in \overline{G}} |u(x) - (P_N u)(x)| \le c \begin{Bmatrix} h^m |\log h|, & r = 1 \\ h^m &, r > 1 \end{Bmatrix}, \tag{7.9}$$

$$\|u - P_N u\|_{L^p(G)} \le c h^m, \qquad r \ge 1,\ 1 \le p < \infty; \tag{7.10}$$

if $m > n-\nu$ then
$$\max_{x \in \overline{G}} |u(x) - (P_N u)(x)| \le c \begin{Bmatrix} h^{r(n-\nu)}, & 1 \le r \le \frac{m}{n-\nu} \\ h^m &, r \ge \frac{m}{n-\nu} \end{Bmatrix}, \tag{7.11}$$

$$\|u - P_N u\|_{L^p(G)} \le c \begin{Bmatrix} h^{r(n-\nu+(1/p))}, & 1 \le r < \frac{m}{n-\nu+(1/p)}, & m > n-\nu+\frac{1}{p} \\ h^m |\log h|^{1/p}, & r = \frac{m}{n-\nu+(1/p)}, & m \ge n-\nu+\frac{1}{p} \\ h^m, & r > \frac{m}{n-\nu+(1/p)}, & r \ge 1 \end{Bmatrix}. \tag{7.12}$$

Proof. (i) Proof scheme. We establish the uniform error bounds on every cell $G_{j_1 \dots j_n}$ separately. Recall that $N = (N_1, \dots, N_n)$. Introduce the interpolation projector P_{N_k} with respect to x_k only:

$$(P_{N_k} u)(x) = \sum_{l=1}^{m} u(x_1, \dots, x_{k-1}, \xi_k^{j_k,l}, x_{k+1}, \dots, x_n) \varphi_k^{j_k,l}(x_k), \quad x \in G_{j_1 \dots j_n},$$

where $\varphi_k^{j_k,l}$ is the polynomial of one variable of degree $m-1$ such that

$$\varphi_k^{j_k,l}(\xi_k^{j_k,l'}) = \begin{Bmatrix} 1, & l = l' \\ 0, & l \ne l' \end{Bmatrix}, \qquad l' = 1, \dots, m. \tag{7.13}$$

On $G_{j_1 \dots j_n}$ we have $P_N u = P_{N_1} P_{N_2} \dots P_{N_n} u$ and

$$u - P_N u = (u - P_{N_1} u) + P_{N_1}(u - P_{N_2} u) + \dots + P_{N_1} P_{N_2} \dots P_{N_{n-1}}(u - P_{N_n} u).$$

It is clear that

$$\|P_{N_k}\|_{\mathcal{L}(C(G_{j_1 \dots j_n}), C(G_{j_1 \dots j_n}))} = \|P\|_{\mathcal{L}(C[-1,1], C[-1,1])}$$

where $P \in \mathcal{L}(C[-1,1], C[-1,1])$ is the standard one dimensional polynomial interpolation projector on $[-1,1]$ corresponding to knots (7.5). Hence

$$\sup_{x\in G_{j_1\ldots j_n}} |u(x)-(P_N u)(x)| \le \|P\|^{n-1} \sum_{k=1}^{n} \sup_{x\in G_{j_1\ldots j_n}} |u(x)-(P_{N_k}u)(x)|. \quad (7.14)$$

In the part (ii) of the proof we show that, for $u\in C_{\square}^{m,\nu}(G)$, $0\le j_k\le N_k-1$,

$$\sup_{x\in G_{j_1\ldots j_n}} |u(x)-(P_{N_k}u)(x)| \le c'(x_k^{j_k+1}-x_k^{j_k})^m \begin{Bmatrix} 1 & , m<n-\nu \\ 1+|\log x_k^{j_k+1}| & , m=n-\nu \\ (x_k^{j_k+1})^{n-\nu-m} & , m>n-\nu \end{Bmatrix}; \quad (7.15)$$

the symmetry argument then yields, for $N_k\le j_k\le 2N_k-1$,

$$\sup_{x\in G_{j_1\ldots j_n}} |u(x)-(P_{N_k}u)(x)| \le c'(x_k^{j_k+1}-x_k^{j_k})^m \begin{Bmatrix} 1 & , m<n-\nu \\ 1+|\log x_k^{j_k}| & , m=n-\nu \\ (x_k^{j_k})^{n-\nu-m} & , m>n-\nu \end{Bmatrix}. \quad (7.15')$$

The constant c' in these inequalities depends on the constant in (7.2) but not on h and the cell. From (7.3), (7.4) and (7.14), (7.15) the uniform error estimates (7.8), (7.9) and (7.11) follow; a detailed argument is given in (iii). Further,

$$\|u-P_N u\|_{L^p(G)} \le \left\{ \sum_{j_1=0}^{2N_1-1} \cdots \sum_{j_n=0}^{2N_n-1} \sup_{x\in G_{j_1\ldots j_n}} |u(x)-(P_N u)(x)|^p \,\mathrm{meas}\, G_{j_1\ldots j_n} \right\}^{1/p} \quad (7.16)$$

and (7.3), (7.4), (7.14), (7.15) yield the L^p estimates (7.10) and (7.12); the details can be found in (iv).

(ii) Proof of (7.15). If v is a polynomial of degree $m-1$ with respect to x_k, with coefficients which may depend on $x_1,\ldots,x_{k-1},x_{k+1},\ldots,x_n$, then $P_{N_k}v=v$, and

$$\sup_{x\in G_{j_1\ldots j_n}} |u(x)-(P_{N_k}u)(x)| \le (1+\|P\|) \sup_{x\in G_{j_1\ldots j_n}} |u(x)-v(x)|.$$

Taking the Taylor polynomial

$$v(x) = \sum_{l=0}^{m-1} \frac{1}{l!} \frac{\partial^l u(x)}{\partial x_k^l}\bigg|_{x_k=x_k^{j_k+1}} (x_k-x_k^{j_k+1})^l$$

we have, due to (7.2),

$$u(x)-v(x) = \frac{1}{(m-1)!} \int_{x_k^{j_k+1}}^{x_k} (x_k-s)^{m-1} \left[\frac{\partial^m u(x)}{\partial x_k^m}\right]_{x_k=s} ds,$$

$$|u(x)-v(x)| \le c \int_{x_k}^{x_k^{j_k+1}} (s-x_k)^{m-1} \begin{Bmatrix} 1 & , & m<n-\nu \\ 1+|\log s|, & & m=n-\nu \\ s^{n-\nu-m} & , & m>n-\nu \end{Bmatrix} ds, \quad x_k^{j_k} \le x_k \le x_k^{j_k+1}.$$

For $1 \le j_k \le N_k-1$ we simply estimate $(s-x_k)^{m-1} \le (x_k^{j_k+1}-x_k^{j_k})^{m-1}$ and $s \ge x_k^{j_k} \ge 2^{-r} x_k^{j_k+1}$; the result is (7.15). For $j_k=0$ we undertake the substitution $x_k = x_k^1 \xi$, $s = x_k^1 \sigma$ transforming the interval $0 \le x_k \le x_k^1$ to interval $0 \le \xi \le 1$. The estimate takes the form

$$|u(x)-v(x)| \le c(x_k^1)^m \begin{Bmatrix} \int_\xi^1 (\sigma-\xi)^{m-1} d\sigma & , & m<n-\nu \\ \int_\xi^1 (\sigma-\xi)^{m-1}(1+|\log x_k^1|+|\log\sigma|)d\sigma, & & m=n-\nu \\ (x_k^1)^{n-\nu-m} \int_\xi^1 (\sigma-\xi)^{m-1}\sigma^{n-\nu-m} d\sigma & , & m>n-\nu \end{Bmatrix}.$$

It is easy to check that

$$c_{m,\mu} := \sup_{0\le\xi\le 1} \int_\xi^1 (\sigma-\xi)^{m-1}\sigma^{-m+\mu} d\sigma < \infty \quad \text{for} \quad \mu>0.$$

The result is (7.15) for $j_k=0$.

(iii) Proof of the uniform errors estimates. Due to the symmetry, it is sufficient to establish the uniform estimates (7.8), (7.9) and (7.11) on any cell $G_{j_1\ldots j_n}$ with $0 \le j_k \le N_k-1$ $(k=1,\ldots,n)$. According to (7.14) and (7.15) we have

$$\sup_{x\in G_{j_1\ldots j_n}} |u(x)-(P_N u)(x)| \le c \sum_{k=1}^n (x_k^{j_k+1}-x_k^{j_k})^m \begin{Bmatrix} 1 & , & m<n-\nu \\ 1+|\log x_k^{j_k+1}| & , & m=n-\nu \\ (x_k^{j_k+1})^{n-\nu-m} & , & m>n-\nu \end{Bmatrix}. \quad (7.17)$$

For $m<n-\nu$ we use (7.4) in the coarse form $x_k^{j_k+1}-x_k^{j_k} \le rb_k/(2N_k) \le rh/2$, and (7.8) immediately follows from (7.17). The same argument yields (7.9) for $m=n-\nu$, $r=1$. To obtain (7.9) for $m=n-\nu$, $r>1$, we take into account that, due to (7.3) and (7.4),

$$(x_k^{j_k+1}-x_k^{j_k})^m(1+|\log x_k^{j_k+1}|) \le cN_k^{-m}\left(\frac{j_k+1}{N_k}\right)^{m(r-1)}\left(1+\left|\log\frac{j_k+1}{N_k}\right|\right) \le c'h^m$$

where the constants are independent of N_k and j_k $(0 \le j_k \le N_k-1)$. Consider the case $m>n-\nu$. Then (7.17) and (7.3), (7.4) yield

$$\begin{aligned} \sup_{x\in G_{j_1\ldots j_n}} |u(x)-(P_N u)(x)| &\le c\sum_{k=1}^n (x_k^{j_k+1}-x_k^{j_k})^m (x_k^{j_k+1})^{n-\nu-m} \\ &\le c'\sum_{k=1}^n \left[(j_k+1)^{r-1}N_k^{-r}\right]^m \left[(j_k+1)^r N_k^{-r}\right]^{n-\nu-m} \\ &= c'\sum_{k=1}^n (j_k+1)^{-m+r(n-\nu)} N_k^{-r(n-\nu)}. \end{aligned}$$

If $r \le m/(n-\nu)$ then $(j_k+1)^{-m+r(n-\nu)} \le 1$ and estimate (7.11) for those r follows. If $r \ge m/(n-\nu)$ then we estimate $j_k+1 \le N_k$, and the result is (7.11) again.

(iv) Proof of $L^p(G)$ estimates. Consider first the case $m=n-\nu$. Making use of the symmetry of estimates (7.15) and (7.15') and estimating $x_k^{j_k+1}-x_k^{j_k}$ by $rh/2$ (see (7.4)) we get from (7.14)-(7.16) estimate (7.10):

$$\|u-P_N u\|_{L^p(G)}$$

$$\le c\left\{\sum_{j_1=0}^{N_1-1}\cdots\sum_{j_n=0}^{N_n-1}\Big[\sum_{k=1}^{n}(x_k^{j_k+1}-x_k^{j_k})^m(1+|\log x_k^{j_k+1}|)\Big]^p \operatorname{meas} G_{j_1\dots j_n}\right\}^{1/p}$$

$$\le c' h^m\left\{\sum_{j_1=0}^{N_1-1}\cdots\sum_{j_n=0}^{N_n-1}\Big[\sum_{k=1}^{n}(1+|\log x_k^{j_k+1}|)\Big]^p \operatorname{meas} G_{j_1\dots j_n}\right\}^{1/p} \le c'' h^m.$$

Let now $m>n-\nu$. We get from (7.14)-(7.16)

$$\|u-P_N u\|_{L^p(G)}$$

$$\le c\left\{\sum_{j_1=0}^{N_1-1}\cdots\sum_{j_n=0}^{N_n-1}\Big[\sum_{k=1}^{n}(x_k^{j_k+1}-x_k^{j_k})^m\,(x_k^{j_k+1})^{n-\nu-m}\Big]^p \operatorname{meas} G_{j_1\dots j_n}\right\}^{1/p}.$$

Summing up over $j_1,\dots,j_{k-1},j_{k+1},\dots,j_n$ this inequality takes the form

$$\|u-P_N u\|_{L^p(G)} \le c'\left\{\sum_{j_k=0}^{N_k-1}\Big[\sum_{k=1}^{n}(x_k^{j_k+1}-x_k^{j_k})^m(x_k^{j_k+1})^{n-\nu-m}\Big]^p(x_k^{j_k+1}-x_k^{j_k})\right\}^{1/p}$$

$$\le c''\left\{\sum_{j_k=0}^{N_k-1}\sum_{k=1}^{n}(x_k^{j_k+1}-x_k^{j_k})^{mp}(x_k^{j_k+1})^{(n-\nu-m)p}\,(x_k^{j_k+1}-x_k^{j_k})\right\}^{1/p}$$

$$\le c'''\sum_{k=1}^{n}\left\{\sum_{j_k=0}^{N_k-1}(x_k^{j_k+1}-x_k^{j_k})^{mp+1}(x_k^{j_k+1})^{(n-\nu-m)p}\right\}^{1/p}.$$

To prove (7.12), it suffices to establish that

$$\left\{\sum_{j_k=0}^{N_k-1}(x_k^{j_k+1}-x_k^{j_k})^{mp+1}(x_k^{j_k+1})^{(n-\nu-m)p}\right\}^{1/p}$$

$$\le c\left\{\begin{array}{ll} h^{r(n-\nu+(1/p))} & ,\quad r<m/(n-\nu+(1/p)) \\ h^{r(n-\nu+(1/p))}(1+|\log h|)^{1/p}, & \quad r=m/(n-\nu+(1/p)) \\ h^m & ,\quad r>m/(n-\nu+(1/p)) \end{array}\right\}. \tag{7.18}$$

Due to (7.3), (7.4) we have

$$\sum_{j_k=0}^{N_k-1}(x_k^{j_k+1}-x_k^{j_k})^{mp+1}(x_k^{j_k+1})^{(n-\nu-m)p}$$

$$\le c\sum_{j_k=0}^{N_k-1}[(j_k+1)^{r-1}N_k^{-r}]^{mp+1}[(j_k+1)^r N_k^{-r}]^{(n-\nu-m)p}$$

$$= c\,N_k^{-r(n-\nu)p-r}\sum_{j_k=0}^{N_k-1}(j_k+1)^{-1+r(n-\nu)p-mp+r}$$

$$\le c' N_k^{-r(n-\nu)p-r}\left\{\begin{array}{ll} 1 & ,\quad r(n-\nu)p-mp+r<0 \\ 1+\log N_k & ,\quad r(n-\nu)p-mp+r=0 \\ N_k^{r(n-\nu)p-mp+r} & ,\quad r(n-\nu)p-mp+r>0 \end{array}\right\},$$

and this is equivalent to (7.18).

The proof of Lemma 7.2 is completed.

Now we present more special error estimates with somewhat reduced values of r.

Lemma 7.3. Let $u\in C_\square^{m,\nu}(G)$ and let $\nu'\in[0,n)$ be a parameter. Then we have, depending on the values of $r\ge 1$:

for $n-\nu'>1$, $\max_{x\in\overline{G}}\int_G |x-y|^{-\nu'}|u(y)-(P_N u)(y)|\,dy$

$$\le c\left\{\begin{array}{ll} h^{r(n-\nu+1)} & ,\quad r<m/(n-\nu+1) \\ h^m(1+|\log h|), & \quad r=m/(n-\nu+1) \\ h^m & ,\quad r>m/(n-\nu+1) \end{array}\right\}; \qquad (7.19)$$

for $n-\nu'\le 1$, $\max_{x\in\overline{G}}\int_G |x-y|^{-\nu'}|u(y)-(P_N u)(y)|\,dy$

$$\le c\left\{\begin{array}{ll} h^{r(2n-\nu-\nu')}(1+|\log h|)^{1-(n-\nu')}, & \quad r<m/(2n-\nu-\nu') \\ h^m(1+|\log h|) & ,\quad r=m/(2n-\nu-\nu') \\ h^m & ,\quad r>m/(2n-\nu-\nu') \end{array}\right\}. \qquad (7.20)$$

Proof. (i) First we prove the inequality

$$\int_0^{b_1}\cdots\int_0^{b_{n-1}}|x-y|^{-\nu}dy_1\ldots dy_{n-1}\le c\left\{\begin{array}{ll} 1 & ,\quad \nu<n-1 \\ 1+|\log|x_n-y_n||, & \quad \nu=n-1 \\ |x_n-y_n|^{n-1-\nu} & ,\quad \nu>n-1 \end{array}\right\}. \qquad (7.21)$$

For $y=(y_1,\ldots,y_n)$ we denote $y'=(y_1,\ldots,y_{n-1})\in\mathbb{R}^{n-1}$ and $a=|x_n-y_n|$. We have

$$a \le \rho + a \le 2a \quad \text{for} \quad \rho \in (0,a),$$

$$\rho \le \rho + a \le 2\rho \quad \text{for} \quad \rho > a,$$

therefore

$$\int_0^{b_1} \ldots \int_0^{b_{n-1}} |x-y|^{-\nu} dy_1 \ldots dy_{n-1} \le c \int_{|x'-y'|<1} (|x'-y'|^2 + a^2)^{-\nu/2} dy' = c \int_{|y'|<1} (|y'|^2 + a^2)^{-\nu/2} dy'$$

$$\sim \int_{|y'|<1} (|y'| + a)^{-\nu} dy' = \sigma_{n-1} \int_0^1 (\rho + a)^{-\nu} \rho^{n-2} d\rho$$

$$\sim \int_0^a a^{-\nu} \rho^{n-2} d\rho + \int_a^1 \rho^{-\nu+n-2} d\rho \sim a^{n-1-\nu} + \left\{ \begin{array}{ll} 1 & , \ \nu < n-1 \\ |\log a| , & \nu = n-1 \\ a^{n-1-\nu}, & \nu > n-1 \end{array} \right\}, \quad n \ge 2,$$

and this yields (7.21).

(ii) Due to (7.17) and symmetries we have

$$v_N(x) := \int_G |x-y|^{-\nu'} |u(y) - (P_N u)(y)| dy \le c \sum_{k=1}^{n} \int_{G'} |x-y|^{-\nu'} \psi_{k,N}(y_k) dy, \quad x \in \overline{G}, \tag{7.22}$$

where (cf. (7.1))

$$G' = \{ y \in \mathbb{R}^n : \ 0 < y_k < b_k/2, \ k = 1, \ldots, n \} \subset G$$

and $\psi_{k,N}$ is a piecewise constant function of one variable,

$$\psi_{k,N}(y_k) = (x_k^{j_k+1} - x_k^{j_k})^m \left\{ \begin{array}{ll} 1 & , \ m < n-\nu \\ 1 + |\log x_k^{j_k+1}| , & m = n-\nu \\ (x_k^{j_k+1})^{n-\nu-m}, & m > n-\nu \end{array} \right\}$$

$$\text{for} \quad x_k^{j_k} < y_k \le x_k^{j_k+1}, \quad j_k = 0,1,\ldots,N_k - 1, \quad k = 1,\ldots,n.$$

Now we integrate in (7.22) the k-th term of the sum with respect to $x_1, \ldots x_{k-1}, x_{k+1}, \ldots, x_n$. On the ground of (7.21),

$$v_N(x) \le c \sum_{k=1}^{n} \int_0^{b_k/2} \left\{ \begin{array}{ll} 1 & , \ \nu' < n-1 \\ 1 + |\log|x_k - y_k|| , & \nu' = n-1 \\ |x_k - y_k|^{n-1-\nu'} & , \ \nu' > n-1 \end{array} \right\} \psi_{k,N}(y_k) dy_k. \tag{7.23}$$

The function $|x_k - y_k|^{n-1-\nu'}$, $\nu' > n-1$, belongs to every $L^q(0, b_k/2)$, $q < 1/(\nu' - n + 1)$; the conjugate p ($p^{-1} + q^{-1} = 1$) satisfies $p > 1/(n-\nu')$. Thus, the Hölder inequality yields

$$\|v_N\|_{C(\bar G)} \le c_p \sum_{k=1}^{n} \left\{ \begin{matrix} \|\psi_{k,N}\|_{L^1(0,b_k/2)}, & \nu'<n-1 \\ \|\psi_{k,N}\|_{L^p(0,b_k/2)}, & \nu'\ge n-1,\ \forall p>1/(n-\nu') \end{matrix} \right\}.$$

Actually, the norm $\|\psi_{k,N}\|_{L^p(0,b_k/2)}$, $1\le p<\infty$, is estimated in the proof of Lemma 7.2 (see e.g. (7.18)):

$$\|\psi_{k,N}\|_{L^p(0,b_k/2)}$$

$$= \Big[\sum_{j_k=0}^{N_k-1} (x_k^{j_k+1} - x_k^{j_k})^{mp} \left\{ \begin{matrix} 1 & , \ \nu<n-1 \\ (1+|\log x_k^{j_k+1}|)^p & , \ \nu=n-1 \\ (x_k^{j_k+1})^{(n-\nu-m)p} & , \ \nu>n-1 \end{matrix} \right\} (x_k^{j_k+1} - x_k^{j_k}) \Big]^{1/p}$$

$$\le c \left\{ \begin{matrix} h^{r(n-\nu+(1/p))} & , \ r<m/(n-\nu+(1/p)) \\ h^m(1+|\log h|)^{1/p}, & r=m/(n-\nu+(1/p)) \\ h^m & , \ r>m/(n-\nu+(1/p)) \end{matrix} \right\}.$$

Now estimate (7.19) follows immediately. Also the third row of (7.20) follows since $r>m/(2n-\nu-\nu')$ implies $r>m/(n-\nu+(1/p))$ with a sufficiently small $p>1/(n-\nu')$. To prove the estimates corresponding to the first two rows of (7.20), we represent the integral in (7.23) as the sum of integrals over the sets $(0,b_k/2)\cap(x_k-h^s,x_k+h^s)$ and $(0,b_k/2)\setminus(x_k-h^s,x_k+h^s)$ where $s>0$ is a parameter. On the first set we integrate the kernel and estimate $\psi_{k,N}$ in L^∞ norm; on the second set we use the Hölder inequality with $q=1/(\nu'+1-n)$ for the kernel and $p=1/(n-\nu')$ for $\psi_{k,N}$ (if $n-\nu'<1$). The result is

$$\|v_N\|_{C(\bar G)} \le c \sum_{k=1}^{n} \Big[h^{s(n-\nu')} \|\psi_{k,N}\|_{L^\infty(0,b_k/2)} + (1+|\log h^s|)^{1/q} \|\psi_{k,N}\|_{L^p(0,b_k/2)} \Big].$$

For $n-\nu'=1$, similar inequality holds. Taking sufficiently great s we see that

$$\|v_N\|_{C(\bar G)} \le c \sum_{k=1}^{n} (1+|\log h|)^{1-(n-\nu')} \|\psi_{k,N}\|_{L^{1/(n-\nu')}(0,b_k/2)}.$$

Now the estimates (7.20), first two rows, follow.

Lemma 7.3 is proved.

7.4. Piecewise polynomial collocation method. Let us denote, for $N=(N_1,\dots,N_n)$, by E_N the finite dimensional space of piecewise polynomial functions u_N on $\bar G$ being polynomials of degree not exceeding $m-1$ with respect to any of arguments $x_1,\dots,x_n$ on any cell $G_{j_1\dots j_n}$ $(0\le j_1\le 2N_1-1,\dots,$ $0\le j_n\le 2N_n-1)$ whereby $u_N\in C(\bar G)$ if $\xi_1=-1$, $\xi_m=1$ (and u_N may have breaks on the plains $x_k = x_k^{j_k}$ ($j_k = 1, \dots, 2N_k - 1$; $k = 1, \dots, n$) if $\xi_1 > -1$ or $\xi_m < 1$. In other words, E_N is the range of the interpolation projector P_N, see Section 7.2.

We look for an approximate solution $u_N \in E_N$ to integral equation (3.1) determining it from the condition of satisfying the equation at collocation points (7.7):

$$\Big[u_N(x) - \int_G K(x,y)u_N(y)dy - f(x)\Big]_{x = (\xi_1^{i_1,l_1}, \dots, \xi_n^{i_n,l_n})} = 0, \tag{7.24}$$

$$l_k = 1, \dots, m, \quad i_k = 0,1, \dots, 2N_k - 1, \quad k = 1, \dots, n.$$

In the case $\xi_1 = -1$, $\xi_m = 1$, a part of collocation points are multiple in $\bar{G}$ — that are the non-corner points with $l_k = 1$ or m for all $k = 1, \dots, n$. It sufficies to collocate at those points only once.

We obtain a linear system of algebraic equations from (7.24) choosing a basis of E_N. For instance, we may use the representation

$$u_N(x) = \sum_{m_1=1}^{m} \dots \sum_{m_n=1}^{m} c_{j_1 \dots j_n}^{m_1 \dots m_n} \varphi_1^{j_1,m_1}(x_1) \dots \varphi_n^{j_n,m_n}(x_n), \quad x \in G_{j_1 \dots j_n}, \tag{7.25}$$

where $\varphi_k^{j_k,m_k}$ are the polymonials of one variable of degree $m-1$ considered on $[x_k^{j_k}, x_k^{j_k+1}]$ and satisfying (7.13). Collocation conditions (7.24) take the form of a system to determine the coefficient $c_{j_1 \dots j_n}^{m_1 \dots m_n}$:

$$c_{i_1 \dots i_n}^{l_1 \dots l_n} = \sum_{j_1=0}^{2N_1-1} \dots \sum_{j_n=0}^{2N_n-1} \sum_{m_1=1}^{m} \dots \sum_{m_n=1}^{m} a_{i_1 \dots i_n j_1 \dots j_n}^{l_1 \dots l_n m_1 \dots m_n} c_{j_1 \dots j_n}^{m_1 \dots m_n}$$
$$+ f(\xi_1^{i_1,l_1}, \dots, \xi_n^{i_n,l_n}), \quad l_k = 1, \dots, m, \quad i_k = 0,1, \dots, 2N_k - 1, \quad k = 1, \dots, n, \tag{7.26}$$

where

$$a_{i_1 \dots i_n j_1 \dots j_n}^{l_1 \dots l_n m_1 \dots m_n}$$
$$= \int_{G_{j_1 \dots j_n}} K(\xi_1^{i_1,l_1}, \dots, \xi_n^{i_n,l_n}, y_1, \dots, y_n) \varphi_1^{j_1,m_1}(y_1) \dots \varphi_n^{j_n,m_n}(y_n) dy_1 \dots dy_n .$$

In the case $\xi_1 = -1$, $\xi_n = 1$, system (7.26) can be somewhat reduced cancelling the equations corresponding to multiple collocations and taking into account the equalities between $c_{j_1 \dots j_n}^{m_1 \dots m_n}$ corresponding to a multiple collocation point.

Independently from the choice of a basis in E_N, collocation conditions (7.24) can be represented as the equation

$$u_N = P_N T u_N + P_N f \tag{7.27}$$

where P_N is the interpolation projector introduced in Section 7.2 and T is the integral operator of equation (3.1).

<u>Theorem</u> 7.1. Let the following conditions be fulfilled:

1. $G \subset \mathbb{R}^n$ is a parallelepided (see (7.1));
2. graded grid (7.3) and the collocation points (7.7) are used;
3. kernel $K(x,y)$ satisfies condition (3.2);
4. $f \in C_{\square}^{m,\nu}(G)$ (see Section 7.1);
5. integral equation (3.1) is uniquely solvable.

Then there are N_k^0 such, for $N_k \ge N_k^0$ $(k=1,\dots,n)$, the collocation conditions (7.24) determine a unique approximation $u_N \in E_N$ to the solution u of equation (3.1). The following error estimates hold:

if $m<n-\nu$ then $\max_{x\in\overline{G}} |u_N(x)-u(x)| \le ch^m$ for any $r \ge 1$;

if $m=n-\nu$ then $\max_{x\in\overline{G}} |u_N(x)-u(x)| \le c \begin{Bmatrix} h^m |\log h|, & r=1 \\ h^m & , r>1 \end{Bmatrix}$;

if $m>n-\nu$ then $\max_{x\in\overline{G}} |u_N(x)-u(x)| \le c \begin{Bmatrix} h^{r(n-\nu)}, & 1 \le r \le m/(n-\nu) \\ h^m & , r \ge m/(n-\nu) \end{Bmatrix}$.

Here $h=\max_{1\le k\le n} b_k/N_k$ and r is the parameter characterizing the non-uniformity of the grid (see (7.3)).

<u>Proof</u>. We consider (3.1) as equation $u=Tu+f$ in the space $E=L^\infty(G)$. According to Lemma 2.2, the integral operator T considered as an operator from $L^\infty(G)$ into $C(\overline{G})$ is compact. This together with the pointwise convergence

$$\|v-P_N v\|_{L^\infty(G)} \to 0 \quad \text{as} \quad N_k \to \infty \quad (k=1,\dots,n), \quad \forall v \in C(\overline{G}),$$

implies the convergence

$$\|T-P_N T\|_{\mathcal{L}(L^\infty(G),L^\infty(G))} \to 0 \quad \text{as} \quad N_k \to \infty \quad (k=1,\dots,n). \tag{7.28}$$

We refer here to Lemma 15.5 in Krasnoselski et al. (1972), but this assertion is elementary for an independent proof too; using Lemma 2.3 it is even possible to give an estimate to $\|T-P_N T\|$. Due to assumption 5, operator $I-T$ has an inverse in $\mathcal{L}(L^\infty(G),L^\infty(G))$, and together with (7.28) we obtain that $I-P_N T$ are for sufficiently large N_k invertible and

$$\|(I-P_N T)^{-1}\|_{\mathcal{L}(L^\infty(G),L^\infty(G))} \le c = \text{const} \quad (N_k \ge N_k^0,\ k=1,\dots,n). \tag{7.29}$$

For the solutions of equations (3.1) and (7.27), u and u_N, we have

$$(I - P_N T)(u_N - u) = P_N f - u + P_N T u = P_N u - u,$$

$$u_N - u = (I - P_N T)^{-1}(P_N u - u), \tag{7.30}$$

$$\|u_N - u\|_{L^\infty(G)} \le c\,\|u - P_N u\|_{L^\infty(G)}, \quad N \ge N_k^0 \quad (k = 1, \dots, n). \tag{7.31}$$

According to Lemma 7.1 $u \in C_\square^{m,\nu}(G)$. Now the error estimates of the Theorem follow from Lemma 7.2. Theorem 7.1 is proved.

Note that an estimate $\sup_{x \in G} |u_N(x) - u(x)| = \mathcal{O}(h^m)$ is of optimal order even for a function $u \in C^\infty(\overline{G})$ and its best approximation in the space of piecewise polynomials of degree $m-1$. Theorem 7.1 shows that, for the collocation method (7.24), the optimal accuracy $\mathcal{O}(h^m)$ can be achieved using sufficiently great values of r. In the practice, too great values of r may cause a numerical instability of the method since, near the boundary ∂G, the grid points will be too close to one another. There are possibilities to reduce r restricting ourselves to $L^p(G)$ estimates of the error $u_N - u$ or to uniform estimates at the collocation points only.

Theorem 7.2. Let the conditions of Theorem 7.1 be fulfilled. Then the error estimate

$$\|u_N - u\|_{L^p(G)} \le c h^m$$

holds for $1 \le p < \infty$ if $\nu < 0$ and for $n/(n-\nu) < p < \infty$ if $\nu \ge 0$, provided that $r = r(m,n,\nu,p)$ is chosen as follows:

$$r \ge 1 \text{ is arbitrary if } m < n - \nu + \frac{1}{p},$$

$$r > \frac{m}{n - \nu + (1/p)} \quad \text{if} \quad m \ge n - \nu + \frac{1}{p}.$$

Proof. We repeat the argument of the proof of Theorem 7.1 with $E = L^p(G)$. Note that for p exposed in the Theorem under the proof, T is compact as an operator from $L^p(G)$ to $C(\overline{G})$; see Lemma 2.2. For those p we obtain the estimate

$$\|u_N - u\|_{L^p(G)} \le c\,\|u - P_N u\|_{L^p(G)}.$$

Now, using estimates (7.10) and (7.12), we obtain the assertion of the Theorem.

7.5. Error estimates at the collocation points. Let us denote

$$\varepsilon_N = \max_{l_k = 1,\dots,m,\ j_k = 0,1,\dots,2N_k - 1,\ k = 1,\dots,n} \left| u_N(x) - u(x) \right|_{x = (\xi_1^{j_1, l_1}, \dots, \xi_n^{j_n, l_n})}. \tag{7.32}$$

This is the maximal error of the approximate solution $u_N \in E_N$ at the collocation points. Since u and $P_N u$ coincide at those points, we have

$$\varepsilon_N = \max_{i_k=1,\dots,m,\, j_k=0,1,\dots,2N_k-1,\, k=1,\dots,n} \left| u_n(x) - (P_N u)(x) \right|_{x=(\xi_1^{j_1,i_1},\dots,\xi_n^{j_n,i_n})}$$

$$\le \|u_N - P_N u\|_{L^\infty(G)}. \tag{7.33}$$

Theorem 7.3. Let the conditions of Theorem 7.1 be fulfilled. Then

$$\varepsilon_N \le c h^m \tag{7.34}$$

provided that $r = r(m,n,\nu) \ge 1$ is restricted by conditions

$$\left.\begin{aligned} r &> \frac{m}{2(n-\nu)} \quad \text{if} \quad 0 < n-\nu \le 1, \\ r &> \frac{m}{n-\nu+1} \quad \text{if} \quad 1 < n-\nu \le m-1 \end{aligned}\right\} \tag{7.35}$$

(and no restriction if $m-1 < n-\nu$).

Proof. For the solutions of equations (3.1) and (7.27) we have

$$(I - P_N T)(u_N - P_N u) = P_N f - P_N u + P_N T P_N u = P_N T P_N u - P_N T u,$$

$$u_N - P_N u = (I - P_N T)^{-1} P_N T (P_N u - u), \tag{7.36}$$

$$\|u_N - P_N u\|_{L^\infty(G)} \le c' \, \|T(u - P_N u)\|_{C(\overline{G})}, \quad N_k \ge N_k^0 \ (k=1,\dots,n), \tag{7.37}$$

where $c' = c\|P_N\|_{\mathcal{L}(C(\overline{G}),L^\infty(G))}$ is independent of N and c is the constant from (7.29). Estimating $\|T(u-P_N u)\|$ according to (7.19), (7.20) with $\nu' = \max\{\nu, 0\}$ we obtain (7.34). Theorem 7.3 is proved.

Remark 7.1. For the iterated approximation

$$\tilde{u}_N(x) = \int_G K(x,y) u_N(y)\,dy + f(x), \quad x \in G, \tag{7.38}$$

we have

$$\max_{x \in \overline{G}} |\tilde{u}_N(x) - u(x)| \le c h^m$$

provided that the conditions of Theorem 7.1 are fulfilled and $r = r(m,n,\nu) \ge 1$ satisfies (7.35).

Indeed, substracting the equality $u = Tu + f$ from (7.38) we get (cf. (7.36) and (7.37)):

$$\tilde{u}_N - u = T(u_N - u) = T(u_N - P_N u) + T(P_N u - u)$$

$$= T(I - P_N T)^{-1} P_N T (P_N u - u) + T(P_N u - u),$$

$$\|\tilde{u}_N - u\|_{C(\overline{G})} \le (\|T\| c \|P_N\| + 1) \|T(u - P_N u)\|_{C(\overline{G})} \le c' h^m.$$

7.6. Superconvergence at the collocation points. Now we assume that $\xi^1, \dots, \xi^m$ (see (7.5)) are the knots of a quadrature formula

$$\int_{-1}^{1} \varphi(\xi)\, d\xi \approx \sum_{l=1}^{m} w_l \varphi(\xi^l), \qquad -1 \le \xi_1 < \xi_2 < \dots < \xi_m \le 1, \tag{7.39}$$

which is sharp for all polynomials of degree $m+\mu$, $0 \le \mu \le m-1$. For instance, the Simpson formula

$$\int_{-1}^{1} \varphi(\xi)\, d\xi \approx \frac{1}{6} [\varphi(-1) + 4\varphi(0) + \varphi(1)], \qquad m = 3,$$

is sharp for cubic polynomials, and our condition is satisfied with $\mu = 0$; the Gauss quadrature formula with m knots $\xi^1, \dots, \xi^m$ (the zeros of the Legendre polynomial of degree m) is sharp for polynomials of degree $2m-1$, i.e. our condition is fulfilled with $\mu = m-1$. Further examples are given by the Lobatto ($\mu = m-3$, $m \ge 3$) and Radau ($\mu = m-2$, $m \ge 2$) quadrature formulas. The Lobatto formula is for us of special interest since $\xi^1 = -1$ and $\xi^m = 1$ in this case; for $m = 3$ the Lobatto and Simpson quadrature formulas coincide.

Using transformation (7.6) we obtain the quadrature formula

$$\int_{x_k^j}^{x_k^{j+1}} \varphi(x_k)\, dx_k \approx \frac{x_k^{j+1} - x_k^j}{2} \sum_{l=1}^{m} w_l \varphi(\xi_k^{j,l}) \tag{7.39'}$$

on $[x_k^j, x_k^{j+1}]$ which remains to be sharp for polynomials of degree $m+\mu$. We introduce the cubature formula

$$\int_{G_{j_1 \dots j_n}} \varphi(x)\, dx \approx 2^{-n} \operatorname{meas} G_{j_1 \dots j_n} \sum_{l_1=1}^{m} \dots \sum_{l_n=1}^{m} w_{l_1} \dots w_{l_n} \varphi(\xi_1^{j_1, l_1}, \dots, \xi_n^{j_n, l_n}). \tag{7.40}$$

It is sharp for all polynomials of degree $m+\mu$ with respect to any of arguments $x_1, \dots, x_n$. Indeed, such a polynomial is a linear combination of terms $x_1^{m_1} \dots x_n^{m_n}$ with $m_k \le m+\mu$ $(k = 1, \dots, n)$ and (7.40) is sharp for them due to the sharpness property of (7.39'):

$$\int_{G_{j_1 \dots j_n}} x_1^{m_1} \dots x_n^{m_n}\, dx = \prod_{k=1}^{n} \int_{x_k^{j_k}}^{x_k^{j_k+1}} x_k^{m_k}\, dx_k = \prod_{k=1}^{n} \frac{x_k^{j_k+1} - x_k^{j_k}}{2} \sum_{l_k=1}^{m} w_{l_k} (\xi_k^{j_k, l_k})^{m_k}$$

$$= 2^{-n} \operatorname{meas} G_{j_1 \dots j_n} \sum_{l_1=1}^{m} \dots \sum_{l_n=1}^{m} w_{l_1} \dots w_{l_n} (\xi_1^{j_1, l_1})^{m_1} \dots (\xi_n^{j_n, l_n})^{m_n}.$$

Let us return to the collocation method (7.24). Recall that ε_N is the maximal error of the method at the collocation points, see (7.32). The following Theorem states convergence rates $\varepsilon_N = O(h^m)$. Thus, the superconvergence phenomenon at collocation points takes place.

Theorem 7.4. Assume that the following conditions are fulfilled:

1. $G \subset \mathbb{R}^n$ is a parallelepided (see (7.1));
2. graded grid (7.3) is used; collocation points (7.7) are generated by the knots $\xi^1, \ldots, \xi^m$ of a quadrature formula (7.39) which is sharp for polynomials of degree $m+\mu$, $0 \le \mu \le m-1$;
3. the kernel $K(x,y)$ is $m+\mu+1$ times continuously differentiable on $(G\times G)\setminus\{x=y\}$ and satisfies (3.2) for $|\alpha|+|\beta| \le m+\mu+1$;
4. $f \in C_{\square}^{m+\mu+1,\nu}(G)$;
5. integral equation (3.1) is uniquely solvable;
6. parameter $r = r(m,n,\nu,\mu) \ge 1$ satisfies the restrictions

$$r > \frac{m}{n-\nu} \quad \text{if} \quad n-\nu < \mu+1; \qquad r \ge \frac{m}{n-\nu} \quad \text{if} \quad n-\nu \ge \mu+1, \qquad r > 1 \quad \text{if} \quad n-\nu = m; \tag{7.41}$$

$$r \ge \frac{m+n-\nu}{n-\nu+1} \quad \text{if} \quad n-\nu < \mu+1; \tag{7.42}$$

$$r > \frac{m+\mu+1}{n-\nu+1} \quad \text{if} \quad n-\nu \ge \mu+1. \tag{7.43}$$

Then

$$\varepsilon_N \le c h^m [\sigma_{n,\nu,\mu}(h) + \tau_{n,\nu}(h)] \tag{7.44}$$

where

$$\sigma_{n,\nu,\mu}(h) = \begin{Bmatrix} h^{\mu+1} & , \; n-\nu > \mu+1 \\ h^{\mu+1}(1+|\log h|), & n-\nu = \mu+1 \\ h^{n-\nu} & , \; n-\nu < \mu+1 \end{Bmatrix}, \quad \tau_{n,\nu}(h) = \begin{Bmatrix} h^n & , \; \nu < 0 \\ h^n(1+|\log h|), & \nu = 0 \\ h^{n-\nu} & , \; \nu > 0 \end{Bmatrix}.$$

In particular, if $\mu+1 \le n$ or if $\nu > 0$ then

$$\varepsilon_N \le c h^m \sigma_{n,\nu,\mu}(h). \tag{7.45}$$

For $\mu+1 > n$, $\nu \le 0$, (7.45) is valid under the following supplementary condition:

7. for $|\alpha| \le \min\{\mu+1-n, -\nu\}$, the function $D_y^\alpha K(x,y)$ is bounded and continuous on $G\times G$ including the diagonal $x=y$.

Note that condition (3.2) guarantees the continuity and boundedness of $D_y^\alpha K(x,y)$, $|\alpha| < \min\{\mu+1-n, -\nu\}$, on $(G\times G)\setminus\{x=y\}$; for $|\alpha| = -\nu$ with $\nu \in \mathbb{Z}$, $|\nu| \le \mu+1-n$, condition 7 strengthens (3.2) banning the (possibly logarithmical) singularity of $D_y^\alpha K(x,y)$ on the diagonal $x=y$; see Remark 7.6.

Proof. According to (7.33) and (7.37),

$$\varepsilon_N \le c \|T(u-P_N u)\|_{C(\bar{G})}. \tag{7.46}$$

We undertake a new estimation of this norm. First we notice that, for $x \notin G_j$,

$$\int_{G_j} K(x,y)[u(y)-(P_N u)(y)]dy = \int_{G_j} [K(x,y)-K_s(x,y)][u(y)-(P_N u)(y)]dy$$

$$+ \sum_{|\alpha|\le s} c_\alpha D_y^\alpha K(x,y^j) \int_{G_j} (y-y^j)^\alpha [u(y)-(P_N^{(\alpha)}u)(y)]dy \tag{7.47}$$

where

$$K_s(x,y) = \sum_{|\alpha|\le s} c_\alpha [D_y^\alpha K(x,y^j)](y-y^j)^\alpha, \quad (y-y^j)^\alpha = (y_1-y_1^{j_1})^{\alpha_1}\dots(y_n-y_n^{j_n})^{\alpha_n},$$

is the Taylor expansion of $K(x,y)$ with respect to y at the center $y^j = (y_1^{j_1}, \dots, y_n^{j_n})$ of the box $G_j = G_{j_1 \dots j_n}$; $c_\alpha = 1/\alpha! = 1/\alpha_1!\dots\alpha_n!$ are the Taylor coefficients; the value of $s \in \mathbb{Z}$ will be chosen later differently in different situations but always $0 \le s \le \mu$; $P_N^{(\alpha)}$ is an interpolation projector similar to P_N but corresponding to the space of piecewise polynomials of degree $m+\mu-|\alpha|$ and the interpolation knots generated by $m+\mu+1-|\alpha|$ knots in $[-1,1]$ — the knots $\xi^1, \dots, \xi^m$ of the quadrature formula (7.39) and additional knots $\xi^{m+1}, \dots, \xi^{m+\mu+1-|\alpha|}$; the choice of last ones in $[-1,1]$ is arbitrary but we assume that they are somehow fixed. To establish (7.47), it suffices to notice that

$$\int_{G_j} (y-y^j)^\alpha [(P_N u)(y)-(P_N^{(\alpha)}u)(y)]dy = 0, \quad |\alpha| \le s \le \mu.$$

This holds due to the sharpness of the cubature formula (7.40) for the polynomials of degree $m+\mu$. Indeed, the integrand is a polynomial of degree not exceeding $m+\mu$ with respect to any of arguments $y_1, \dots, y_n$ and it vanishes at knots (7.7), therefore the integral is equal to its cubature approximation and the cubature approximation is zero.

Summing up over the boxes G_j we obtain from (7.47)

$$\int_G K(x,y)[u(y)-(P_N u)(y)]dy = \sum_{G_j \cap B(x,h) \ne \emptyset} \int_{G_j} K(x,y)[u(y)-(P_N u)(y)]dy$$

$$+ \sum_{G_j \cap B(x,h) = \emptyset} \int_{G_j} [K(x,y)-K_s(x,y)][u(y)-(P_N u)(y)]dy$$

$$+ \sum_{G_j \cap B(x,h) = \emptyset} \sum_{|\alpha|\le s} c_\alpha \int_{G_j} [D_y^\alpha K(x,y^j)](y-y^j)^\alpha [u(y)-(P_N^{(\alpha)}u)(y)]dy. \tag{7.48}$$

Due to (3.3),

$$\sum_{G_j \cap B(x,h) \ne \emptyset} \int_{G_j} |K(x,y)|dy \le c\tau_{n,\nu}(h) \tag{7.49}$$

with $\tau_{n,\nu}$ defined after (7.44). Further, due to (3.2), we have for $y \in G_j$ with $G_j \cap B(x,h) = \emptyset$

$$|D_y^\alpha K(x,y^j)| \le b \begin{Bmatrix} 1 & , & \nu+|\alpha|<0 \\ 1+|\log|x-y^j||, & & \nu+|\alpha|=0 \\ |x-y^j|^{-\nu-|\alpha|} & , & \nu+|\alpha|>0 \end{Bmatrix} \le c \begin{Bmatrix} 1 & , & \nu+|\alpha|<0 \\ 1+|\log|x-y||, & & \nu+|\alpha|=0 \\ |x-y|^{-\nu-|\alpha|} & , & \nu+|\alpha|>0 \end{Bmatrix},$$

$$|K(x,y) - K_s(x,y)|$$

$$\le ch^{s+1} \max_{z \in G_j} \begin{Bmatrix} 1 & , & \nu+s+1<0 \\ 1+|\log|x-z||, & & \nu+s+1=0 \\ |x-z|^{-\nu-s-1} & , & \nu+s+1>0 \end{Bmatrix} \le c'h^{s+1} \begin{Bmatrix} 1 & , & \nu+s+1<0 \\ 1+|\log|x-y||, & & \nu+s+1=0 \\ |x-y|^{-\nu-s-1} & , & \nu+s+1>0 \end{Bmatrix},$$

therefore (see (3.4))

$$\int_{G \setminus B(x,h)} |K(x,y)-K_s(x,y)|\,dy \le ch^{s+1} \begin{Bmatrix} 1 & , & \nu+s+1<n \\ 1+|\log h| & , & \nu+s+1=n \\ h^{n-\nu-s-1} & , & \nu+s+1>n \end{Bmatrix}$$

$$= c \begin{Bmatrix} h^{s+1} & , & n-\nu>s+1 \\ h^{s+1}(1+|\log h|), & & n-\nu=s+1 \\ h^{n-\nu} & , & n-\nu<s+1 \end{Bmatrix}.$$

We get from (7.46) and (7.48)

$$\varepsilon_N \le c_1 \|u-P_N u\|_{L^\infty(G)} \tau_{n,\nu} + c_2 \|u-P_N u\|_{L^\infty(G)} \begin{Bmatrix} h^{s+1} & , & n-\nu>s+1 \\ h^{s+1}(1+|\log h|), & & n-\nu=s+1 \\ h^{n-\nu} & , & n-\nu<s+1 \end{Bmatrix}$$

$$+ c_3 \sum_{|\alpha| \le s} h^{|\alpha|} \sup_{x \in G} \int_{G \setminus B(x,h)} \begin{Bmatrix} 1 & , & \nu+|\alpha|<0 \\ 1+|\log|x-y||, & & \nu+|\alpha|=0 \\ |x-y|^{-\nu-|\alpha|} & , & \nu+|\alpha|>0 \end{Bmatrix} |u(y)-(P_N^{(\alpha)}u)(y)|\,dy. \quad (7.50)$$

We put $s=\mu$ if $n-\nu \ge \mu+1$ and $s=[n-\nu]$, the integer part of $n-\nu$, if $n-\nu<\mu+1$.

Due to (7.41) we have (see (7.9) and (7.11)) $\|u-P_N u\|_{L^\infty(G)} \le ch^m$, and the first two terms on the right hand side of (7.50) fit into the error estimate (7.44). Let us prove the same for the integral terms:

$$h^{|\alpha|} \sup_{x \in G} \int_{G \setminus B(x,h)} \begin{Bmatrix} 1 & , & \nu+|\alpha|<0 \\ 1+|\log|x-y||, & & \nu+|\alpha|=0 \\ |x-y|^{-\nu-|\alpha|} & , & \nu+|\alpha|>0 \end{Bmatrix} |u(y)-(P_N^{(\alpha)}u)(y)|\,dy$$

$$\le ch^m \begin{Bmatrix} h^{\mu+1} & , & n-\nu>\mu+1 \\ h^{\mu+1}(1+|\log h|), & & n-\nu=\mu+1 \\ h^{n-\nu} & , & n-\nu<\mu+1 \end{Bmatrix} \quad (0\le|\alpha|\le s). \tag{7.51}$$

We make use of Lemma 7.3 replacing there m by $m+\mu+1-|\alpha|$ (the degree of piecewise polynomials in the range of $P_N^{(\alpha)}$ is $m+\mu-|\alpha|$, not $m-1$ as for P_N in Lemma 7.3). Assume that $\nu' = \nu+|\alpha|>0$. Inequalities (7.19) and (7.20) coincide for $n-\nu'=1$ and take the following form:

for $n-(\nu+|\alpha|)\ge 1$, $\quad \sup\limits_{x\in G} \int\limits_G |x-y|^{-\nu-|\alpha|}|u(y)-(P_N^{(\alpha)}u)(y)|dy$

$$\le c \begin{Bmatrix} h^{r(n-\nu+1)} & , & r<(m+\mu+1-|\alpha|)/(n-\nu+1) \\ h^{m+\mu+1-|\alpha|}(1+|\log h|), & & r=(m+\mu+1-|\alpha|)/(n-\nu+1) \\ h^{m+\mu+1-|\alpha|} & , & r>(m+\mu+1-|\alpha|)/(n-\nu+1) \end{Bmatrix}, \tag{7.52}$$

for $0<n-(\nu+|\alpha|)\le 1$, $\quad \sup\limits_{x\in G} \int\limits_G |x-y|^{-\nu-|\alpha|}|u(y)-(P_N^{(\alpha)}u)(y)|dy$

$$\le c \begin{Bmatrix} h^{r(2n-2\nu-|\alpha|)}(1+|\log h|)^{1-(n-(\nu+|\alpha|))}, & r<(m+\mu+1-|\alpha|)/(2n-2\nu-|\alpha|) \\ h^{m+\mu+1-|\alpha|}(1+|\log h|) \quad , & r=(m+\mu+1-|\alpha|)/(2n-2\nu-|\alpha|) \\ h^{m+\mu+1-|\alpha|} \quad , & r>(m+\mu+1-|\alpha|)/(2n-2\nu-|\alpha|) \end{Bmatrix}. \tag{7.53}$$

Consider first the case $n-\nu\ge\mu+1$, $s=\mu$. Then $n-(\nu+|\alpha|)\ge n-(\nu+s)\ge 1$, and the third row of (7.52) yields (7.51):

$$h^{|\alpha|}\sup_{x\in G}\int_G |x-y|^{-\nu-|\alpha|}|u(y)-(P_N^{(\alpha)}u)(y)|dy \le ch^{m+\mu+1}, \quad 0\le|\alpha|\le s,$$

provided that $r>(m+\mu+1-|\alpha|)/(n-\nu+1)$, $0\le|\alpha|\le s$; due to (7.43), these conditions on r are satisfied. It is easy to analyse the cases $\nu+|\alpha|<0$ and $\nu+|\alpha|=0$, too, and we always get (7.51). Now we consider the case $n-\nu<\mu+1$, $s=[n-\nu]$. If $\nu+|\alpha|\le n-1$ then, for r restricted by (7.42), we obtain from (7.52) at least

$$h^{|\alpha|}\sup_{x\in G}\int_G |x-y|^{-\nu-|\alpha|}|u(y)-(P_N^{(\alpha)}u)(y)|dy \le ch^{m+(n-\nu)}. \tag{7.54}$$

In other words, (7.51) is valid for $|\alpha|<s=[n-\nu]$. But there are also terms with $|\alpha|=[n-\nu]$ whereby $n-1<\nu+|\alpha|<n$ if $\nu\notin\mathbb{Z}$ and $\nu+|\alpha|=n$ if $\nu\in\mathbb{Z}$. In the case $\nu\notin\mathbb{Z}$ we get (7.54) from (7.53) provided that $r(2n-2\nu)>m+(n-\nu)$, and this condition on $r\ge 1$ is satisfied due to (7.41). In the case $\nu\in\mathbb{Z}$ the corresponding terms in (7.50) are of the form

$$h^{n-\nu} \sup_{x\in G} \int_{G\setminus B(x,h)} |x-y|^{-n} |u(y)-(P_N^{(\alpha)}u)(y)|dy, \quad |\alpha| = n-\nu;$$

according to (7.11) we have

$$\max_{y\in \overline{G}} |u(y)-(P_N^{(\alpha)}u)(y)| \le c \left\{ \begin{array}{ll} h^{r(n-\nu)} & , \ 1\le r\le (m+\mu+1-(n-\nu))/(n-\nu) \\ h^{m+\mu+1-(n-\nu)}, & r\ge (m+\mu+1-(n-\nu))/(n-\nu) \end{array} \right\},$$

and taking into account (3.4) we again obtain (7.54) for $r>m/(n-\nu)$. Thus, (7.51) holds for all terms with $|\alpha|\le s=[n-\nu]$. We have finished the proof of estimate (7.44).

Now we turn to the proof of estimate (7.45) for $\mu+1>n$, $\nu\le 0$ under supplementary condition 7 of the Theorem. We only must find a better estimation to the terms of the first sum on the right hand side of equality (7.48). The case $\nu=0$ is clear. Let $\nu>0$. The idea is to apply a representation of type (7.47) for those terms too using now the point x as the center of the Taylor expansion:

$$\int_{G_j} K(x,y)[u(y)-(P_N u)(y)]dy = \int_{G_j} [K(x,y)-K_s(x,y)][u(y)-(P_N u)(y)]dy$$

$$+ \sum_{|\alpha|\le s} c_\alpha D_y^\alpha K(x,y)\Big|_{y=x} \int_{G_j} (y-x)^\alpha [u(y)-(P_N^{(\alpha)}u)(y)]dy, \qquad (7.55)$$

$$K_s(x,y) = \sum_{|\alpha|\le s} c_\alpha D_y^\alpha K(x,y)\Big|_{y=x} (y-x)^\alpha, \qquad y\in G_j,\ G_j\cap B(x,h)\ne\emptyset.$$

We put (i) $s=\mu-n$ if $|\nu|>\mu+1-n$; (ii) $s=|\nu|-1$ if $|\nu|\le\mu+1-n$, $\nu\in\mathbb{Z}$; (iii) $s=[|\nu|]$ if $|\nu|\le\mu+1-n$, $\nu\notin\mathbb{Z}$. Then, respectively, (i) $\nu+s+1<0$, (ii) $\nu+s+1=0$, (iii) $0<\nu+s+1<1$. The remainder of the Taylor expansion can be represented in the form (see e.g. Dieudonné (1960))

$$K(x,y)-K_s(x,y) = \int_0^1 \frac{(1-\xi)^s}{s!} K^{(0,s+1)}(x, x+\xi(y-x))\,d\xi\,(y-x)^{(s+1)}$$

where $K^{(0,s+1)}(x,y)$ is the Fréchet derivative of order s+1 of $K(x,y)$ as a function of y. According to (3.2') and condition 7 of the Theorem, we get

$$|K(x,y)-K_s(x,y)| \le c \left\{ \begin{array}{ll} 1 & , \text{ cases (i) and (ii)} \\ |x-y|^{-\nu-(s+1)}, & \text{case (iii)} \end{array} \right\} |y-x|^{s+1}$$

$$= c \left\{ \begin{array}{ll} |x-y|^{s+1}, & \text{cases (i) and (ii)} \\ |x-y|^{-\nu} \ , & \text{case (iii)} \end{array} \right\} = \left\{ \begin{array}{ll} |x-y|^{\mu+1-n}, & \text{case (i)} \\ |x-y|^{-\nu} & , \text{ cases (ii) and (iii)} \end{array} \right\}.$$

Together with $\|u-P_N u\|_{L^\infty(G)} \le ch^m$ we get

$$\sum_{G_j \cap B(x,h) \neq \emptyset} \left| \int_{G_j} [K(x,y)-K_s(x,y)][u(y)-(P_N u)(y)]dy \right|$$

$$\le ch^m h^n \begin{Bmatrix} h^{\mu+1-n}, & \text{case (i)} \\ h^{-\nu} & , \text{cases (ii) and (iii)} \end{Bmatrix} \le ch^m h^{\min\{\mu+1,\, n-\nu\}},$$

and this supports (7.45). The terms of the second sum of (7.55) can be easily estimated in a similar manner as above, cf. (7.51), even neglecting the fact that now the domain of integration G_j is small. Note that $|D_y^\alpha K(x,y)| \le \text{const}$ for $|\alpha| \le s$.

The proof of Theorem 7.4 is completed.

Remark 7.2. Under conditions 1-6 of Theorem 7.4,

$$\max_{y \in \overline{G}} |\tilde{u}_N(x)-u(x)| \le ch^m[\sigma_{n,\nu,\mu}(h)+\tau_{n,\nu}(h)];$$

under conditions 1-7,

$$\max_{y \in \overline{G}} |\tilde{u}_N(x)-u(x)| \le ch^m \sigma_{n,\nu,\mu}(h)$$

where u_N is defined by (7.38) and $\sigma_{n,\nu,\mu}$, $\tau_{n,\nu}$ by (7.44).

Remark 7.3. In the case $\mu = 0$ conditions (7.41) - (7.43) reduce to (7.41):

$$r > \frac{m}{n-\nu} \quad \text{if } n-\nu < 1; \quad r \ge \frac{m}{n-\nu} \quad \text{if } n-\nu \ge 1, \quad r > 1 \quad \text{if } n-\nu = m;$$

estimate (7.44) takes the form

$$\varepsilon_N \le ch^m \begin{Bmatrix} h & , n-\nu > 1 \\ h(1+|\log h|), & n-\nu = 1 \\ h^{n-\nu} & , n-\nu < 1 \end{Bmatrix}.$$

Remark 7.4. In the case $\mu = m-1$ (the Gauss quadrature formula) conditions (7.41) - (7.43) reduce to

$$r > \frac{m}{n-\nu}, \quad n-\nu \le \sqrt{m},$$

$$r \ge \frac{m+n-\nu}{n-\nu+1}, \quad \sqrt{m} < n-\nu < m,$$

$$r > \frac{2m}{n-\nu+1}, \quad n-\nu \ge m;$$

estimate (7.44) takes the form

$$\varepsilon_N \le c_1 h^m \left\{ \begin{array}{ll} h^m & , \ n-\nu > m \\ h^m(1+|\log h|), & n-\nu = m \\ h^{n-\nu} & , \ n-\nu < m \end{array} \right\} + c_2 h^m \left\{ \begin{array}{ll} h^n & , \ \nu < 0 \\ h^n(1+|\log h|), & \nu = 0 \\ h^{n-\nu} & , \ \nu > 0 \end{array} \right\};$$

estimate (7.45) takes the form

$$\varepsilon_n = ch^m \left\{ \begin{array}{ll} h^m & , \ n-\nu > m \\ h^m(1+|\log h|), & n-\nu = m \\ h^{n-\nu} & , \ n-\nu < m \end{array} \right\}.$$

<u>Remark</u> 7.5. If $n-\nu > m+\mu$ then conditions (7.41) - (7.43) are satisfied for any $r \ge 1$; estimate (7.45) takes the form $\varepsilon_N \le ch^{m+\mu+1}$.

<u>Remark</u> 7.6. Let us weaken for $\mu+1 > n$, $\nu \le 0$ condition 7 of Theorem 7.4 as follows: for $|\alpha| < \min\{\mu+1-n \ , \ -\nu\}$, the functions $D_y^\alpha K(x,y)$ are bounded and continuous on $G \times G$ including the diagonal $x = y$. Then (cf. (7.45))

$$\varepsilon_N \le ch^m \cdot \left\{ \begin{array}{ll} h^{\mu+1} & , \ n-\nu > \mu+1 \\ h^{n-\nu}(1+|\log h|), & n-\nu \le \mu+1, \ \nu \in \mathbb{Z} \\ h^{n-\nu} & , \ n-\nu \le \mu+1, \ \nu \notin \mathbb{Z} \end{array} \right\}.$$

Now we illustrate Theorems 7.1, 7.2 and 7.4 for $m = 1,2,3$.

7.7. Piecewise constant collocation ($m = 1$). In the case $m = 1$, $\xi^1 = 0$ is the most suitable choice of the single point $\xi^1 \in [-1,1]$. The quadrature formula $\int_{-1}^{1} \varphi(\xi)\, d\xi \approx 2\varphi(0)$ is sharp for polynomials of degree 1, and we may apply Theorem 7.4 with $\mu = 0$. Together with Theorem 7.1 we have the following error estimates:

$$\max_{y \in \overline{G}} |u_N(x) - u(x)| \le ch \quad \text{for} \quad \left\{ \begin{array}{lll} r \ge 1/(n-\nu) & \text{if} & n-\nu < 1 \\ r > 1 & \text{if} & n-\nu = 1 \\ r \ge 1 & \text{if} & n-\nu > 1 \end{array} \right\},$$

$$\varepsilon_N \le c \left\{ \begin{array}{ll} h^2 & \text{for } n-\nu > 1, \ r \ge 1 \\ h^2(1+|\log h|) & \text{for } n-\nu = 1, \ r > 1 \\ h^{1+n-\nu} & \text{for } n-\nu < 1, \ r > 1/(n-\nu) \end{array} \right\}. \tag{7.56}$$

This simplest collocation method (PCCM) is carefully examined in Section 5 for more general domains $G \subset \mathbb{R}^n$. According to Theorem 5.1, estimate (5.17),

$$\varepsilon_N \le c \left\{ \begin{array}{ll} h^2 & , \ n-\nu > 1 \\ h^2(1+|\log h|)^2, & n-\nu = 1 \\ h^{2(n-\nu)} & , \ n-\nu < 1 \end{array} \right\}. \tag{7.57}$$

This result corresponds to $r=1$ since no scaling of the grid was used in Section 5. Estimates (7.56) and (7.57) coincide for $\nu<n-1$; using graded grids with $r>1/(n-\nu)$ we have achieved a slight improvement for $\nu=n-1$ and a more significant improvement for $\nu>n-1$. This improvement corresponds to a better approximation of the solution of integral equation (3.1) by piecewise constant functions on non-uniform grids. There remains a gap between the convergence rates $\mathcal{O}(h^2)$ for $\nu<n-1$ and $\mathcal{O}(h^{1+n-\nu})$ for $\nu>n-1$. The smaller convergence rate $\mathcal{O}(h^{1+n-\nu})$ in the case $\nu>n-1$ is caused by the more strong singularity of the kernel and cannot be removed by the refinement of the grid near the boundary only.

7.8. Piecewise polylinear collocation (m=2). In the case $m=2$, there are two choice of the points $\xi^1, \xi^2 \in [-1,1]$ of an interest.

(i) $\xi^1 = -1$, $\xi^2 = 1$. In this case the approximate solution $u_N \in E_N$ can be represented in the form of a continuous piecewise polylinear function

$$u_N(x) = \sum_{j_1=0}^{2N_1} \cdots \sum_{j_n=0}^{2N_n} c_{j_1 \dots j_n} \varphi_1^{j_1}(x_1) \dots \varphi_n^{j_n}(x_n), \quad x \in \overline{G},$$

where $\varphi_k^i(x_k)$, $i=0,1,\dots,2N_k$, are one dimensional basic linear splines corresponding to grid points (7.3), i.e. $\varphi_k^i(x_k)$ is linear on every interval $[x_k^j, x_k^{j+1}]$, $j=0,1,\dots,2N_k-1$, $\varphi_k^i(x_k^j) = 0$ for $i \neq j$ and $\varphi_k^i(x_k^i) = 1$. Collocation method (7.24) leads to the system

$$c_{i_1 \dots i_n} = \sum_{j_1=0}^{2N_1} \cdots \sum_{j_n=0}^{2N_n} a_{i_1 \dots i_n j_1 \dots j_n} c_{j_1 \dots j_n} + f(x_1^{i_1}, \dots, x_n^{i_n}),$$

$$i_k = 0,1,\dots,2N_k, \quad k=1,\dots,n,$$

where

$$a_{i_1 \dots i_n j_1 \dots j_n} = \int_G K(x_1^{i_1}, \dots, x_n^{i_n}, y_1, \dots, y_n)\, \varphi_1^{j_1}(y_1) \dots \varphi_n^{j_n}(y_n) dy_1 \dots dy_n.$$

There is no quadrature formula $\int_{-1}^{1} \varphi(\xi)\, d\xi \approx w_{-1}\varphi(-1) + w_1\varphi(1)$ which is sharp for polynomials of degree 2. Thus, Theorem 7.4 may not be applied. From Theorems 7.1 and 7.3 we obtain the following error estimates:

$$\max_{y \in \overline{G}} |u_N(x) - u(x)| \le ch^2 \quad \text{for} \quad \left\{ \begin{array}{lll} r \ge 2/(n-\nu) & \text{if} & n-\nu<2 \\ r>1 & \text{if} & n-\nu=2 \\ r \ge 1 & \text{if} & n-\nu>2 \end{array} \right\}, \tag{7.58}$$

$$\varepsilon_N \le ch^2 \quad \text{for} \quad \left\{ \begin{array}{lll} r>1/(n-\nu) & \text{if} & n-\nu \le 1 \\ r \ge 1 & \text{if} & n-\nu>1 \end{array} \right\}. \tag{7.59}$$

(ii) $-\xi^1 = \xi^2 = 1/\sqrt{3}$ (ξ^1 and ξ^2 are knots of the Gauss quadrature formula, $m=2$). The representation (7.25), (7.26) of the collocation method seems to be most suitable. From Theorems 7.1, 7.3 and 7.4, $\mu=1$, we obtain estimates (7.58), (7.59) and the estimate

$$\varepsilon_N \le ch^2 \begin{Bmatrix} h^2 & , n-\nu > 2 \\ h^2(1+|\log h|)^2, & n-\nu = 2 \\ h^{n-\nu} & , n-\nu < 2 \end{Bmatrix} \text{ for } \begin{Bmatrix} r > 2/(n-\nu) & \text{if} & n-\nu \le \sqrt{2} \\ r \ge (2+n-\nu)/(1+n-\nu) & \text{if} & \sqrt{2} < n-\nu < 2 \\ r > 4/(1+n-\nu) & \text{if} & 2 \le n-\nu \le 3 \\ r \ge 1 & \text{if} & n-\nu > 3 \end{Bmatrix}. \quad (7.60)$$

7.9. Piecewise polysquare collocation (m=3). Again, two choices of the points $\xi^1, \xi^2, \xi^3 \in [-1,1]$ are of special interest.

(i) $\xi^1 = -1$, $\xi^2 = 0$, $\xi^3 = 1$ with Simpson quadrature formula (see Section 7.6). Theorem 7.4 may be applied with $\mu = 0$. From Theorems 7.1, 7.2 and 7.4 we obtain the error estimates

$$\max_{y \in \bar{G}} |u_N(x) - u(x)| \le ch^3 \quad \text{for} \quad \begin{Bmatrix} r \ge 3/(n-\nu) & \text{if} & n-\nu < 3 \\ r > 1 & \text{if} & n-\nu = 3 \\ r \ge 1 & \text{if} & n-\nu > 3 \end{Bmatrix}, \quad (7.61)$$

$$\varepsilon_N \le ch^3 \quad \text{for} \quad \begin{Bmatrix} r > 3/(2(n-\nu)) & \text{if} & n-\nu \le 1 \\ r > 3/(n-\nu+1) & \text{if} & 1 < n-\nu \le 2 \\ r \ge 1 & \text{if} & n-\nu > 2 \end{Bmatrix}, \quad (7.62)$$

$$\varepsilon_N \le ch^3 \begin{Bmatrix} h & , n-\nu > 1 \\ h(1+|\log h|), & n-\nu = 1 \\ h^{n-\nu} & , n-\nu < 1 \end{Bmatrix} \quad \text{for} \quad \begin{Bmatrix} r > 3/(n-\nu) & \text{if} & n-\nu < 1 \\ r \ge 3/(n-\nu) & \text{if} & 1 \le n-\nu < 3 \\ r > 1 & \text{if} & n-\nu = 3 \\ r \ge 1 & \text{if} & n-\nu > 3 \end{Bmatrix}. \quad (7.63)$$

(ii) $-\xi^1 = \xi^3 = \sqrt{3/5}$, $\xi^2 = 0$ (ξ^1, ξ^2 and ξ^3 are the knots of the Gauss quadrature formula, $m=3$). Estimates (7.61) - (7.63) hold again. Theorem 7.4 may be applied with $\mu=2$, too, and we obtain in the case $\{n=2,\ \nu>0\}$ and $n \ge 3$ the estimate

$$\varepsilon_N \le ch^3 \begin{Bmatrix} h^3 & , n-\nu > 3 \\ h^3(1+|\log h|), & n-\nu = 3 \\ h^{n-\nu} & , n-\nu < 3 \end{Bmatrix} \text{ for } \begin{Bmatrix} r > 3/(n-\nu) & \text{if} & n-\nu \le \sqrt{3} \\ r \ge (3+n-\nu)/(1+n-\nu) & \text{if} & \sqrt{3} < n-\nu < 3 \\ r > 6/(1+n-\nu) & \text{if} & 3 \le n-\nu \le 5 \\ r \ge 1 & \text{if} & n-\nu > 5 \end{Bmatrix}. \quad (7.64)$$

In the case $n=2$, $\nu<0$ estimate (7.64) holds provided that condition 7 of Theorem 7.4 is satisfied.

7.10. Exercises. 7.10.1. Assume (3.2). Using Lemma 2.3 prove that

$$\|T - P_N T\|_{\mathcal{L}(C(\overline{G}),C(\overline{G}))} \le c \begin{Bmatrix} h & , & \nu < n-1 \\ h(1+|\log h|), & & \nu = n-1 \\ h^{n-\nu} & , & \nu > n-1 \end{Bmatrix}.$$

Propose estimates for N_k^o $(k=1,\dots,n)$ in Theorem 7.1 provided that an estimate $\|(I-T)^{-1}\|_{\mathcal{L}(C(\overline{G}),C(\overline{G}))} \le c_0$ is known.

7.10.2. Complete Theorem 7.2 with estimates of $\|u_N - u\|_{L^p(G)}$ for r between 1 and $m/(n-\nu+(1/p))$.

7.10.3. Complete Theorem 7.3 with estimates of ε_N for $r \ge 1$ not satisfying (7.35).

7.10.4. Complete Theorem 7.4 with estimates of ε_N for $r \ge 1$ not satisfying (7.41)-(7.43).

7.10.5. Prove Remark 7.6.

7.10.6. Estimate $\|u_N - u\|_{L^p(G)}$ for the methods considered in Sections 7.7-7.9.

7.10.7. Illustrate Theorems 7.1-7.4 for $m=4$ using (i) Lobatto knots; (ii) Gauss knots.

8. NONLINEAR INTEGRAL EQUATION

In this Chapter we generalize some of the main results of the previous chapters to the case of nonlinear integral equations. Giving extensive formulations of the results and the proof ideas, we only outline the technical details of the proofs which are similar to the case of linear integral equations.

8.1. Smoothness of the solution. Consider the integral equation

$$u(x) = \int_G \mathcal{K}(x,y,u(y))dy + f(x), \quad x \in G, \tag{8.1}$$

where $G \subset \mathbb{R}^n$ is an open bounded set. The kernel $\mathcal{K}(x,y,u)$ is assumed to be m times ($m \ge 1$) continuously differentiable with respect to x, y, u for $x \in G$, $y \in G$, $x \ne y$, $u \in (-\infty,\infty)$ whereby there exists a real number $\nu \in (-\infty,n)$ such that, for any $k \in \mathbb{Z}_+$ and $\alpha \in \mathbb{Z}^n_+$, $\beta \in \mathbb{Z}^n_+$ with $k + |\alpha| + |\beta| \le m$, the following inequalities hold:

$$\left| D^\alpha_x D^\beta_{x+y} \frac{\partial^k}{\partial u^k} \mathcal{K}(x,y,u)\right| \le b_1(u) \begin{Bmatrix} 1 & , & \nu+|\alpha|<0 \\ 1+|\log|x-y||, & & \nu+|\alpha|=0 \\ |x-y|^{-\nu-|\alpha|} & , & \nu+|\alpha|>0 \end{Bmatrix}, \tag{8.2}$$

$$\left| D^\alpha_x D^\beta_{x+y} \frac{\partial^k}{\partial u^k} \mathcal{K}(x,y,u_1) - D^\alpha_x D^\beta_{x+y} \frac{\partial^k}{\partial u^k} \mathcal{K}(x,y,u_2)\right|$$

$$\le b_2(u_1,u_2)\,|u_1-u_2| \begin{Bmatrix} 1 & , & \nu+|\alpha|<0 \\ 1+|\log|x-y||, & & \nu+|\alpha|=0 \\ |x-y|^{-\nu-|\alpha|} & , & \nu+|\alpha|>0 \end{Bmatrix}. \tag{8.3}$$

The functions $b_1: \mathbb{R} \to \mathbb{R}_+$ and $b_2: \mathbb{R}^2 \to \mathbb{R}_+$ are assumed to be bounded on every bounded region of $\mathbb{R}^1$ and $\mathbb{R}^2$, respectively. Other designations are explained in Section 3.1. Note that in the case of linear equation ($\mathcal{K}(x,y,u) = K(x,y)u$), conditions (8.2) and (8.3) reduce to (3.2). We refer to Section 2.5 and 2.6 for the definitions of the weighted space $C^{m,\nu}(G)$ and $C_a^{m,\nu}(G)$.

<u>Theorem</u> 8.1. Let $f \in C^{m,\nu}(G)$ and let the kernel $\mathcal{K}(x,y,u)$ satisfy conditions (8.2) and (8.3). If integral equation (8.1) has a solution $u \in L^\infty(G)$

then $u \in C^{m,\nu}(G)$.

<u>Theorem</u> 8.2. Assume that the boundary ∂G is piecewise smooth and $a(x) = (a_1(x), \dots, a_n(x))$, $x \in \bar{G}$, is a vector field of the class $[C^m(\bar{G})]^n$. Let $f \in C_a^{m,\nu}(G)$ and let the kernel $\mathcal{K}(x,y,u)$ satisfy (8.2) and (8.3). If integral equation (8.1) has a solution $u \in L^\infty(G)$ then $u \in C_a^{m,\nu}(G)$.

Proof. The proof of Theorems 8.1 and 8.2 is based on the same ideas as the proof of Theorems 3.1 and 3.2 in Sections 3.5 and 3.8. Let $u_0 \in L^\infty(G)$ be a solution to (8.1). We present equation (8.1) in the form

$$u(x) = \int_\Omega \mathcal{K}(x,y,u(y))\,dy + f_\Omega(x), \quad x \in \Omega = G \cap B(\bar{x},\delta), \tag{8.4}$$

where

$$f_\Omega(x) = f(x) + \int_{G \setminus \Omega} \mathcal{K}(x,y,u_0(y))\,dy, \quad x \in \Omega; \tag{8.5}$$

the ball $B(\bar{x},\delta) \subset \mathbb{R}^n$ has an arbitrary center $\bar{x} \in G$ and a small radius $\delta > 0$, $\delta = \min\{\delta_0, \rho(\bar{x})/2\}$ in the case of Theorem 8.1 and $\delta = \min\{\delta_0, \rho_a(\bar{x})/2\}$ in the case of Theorem 8.2. Differentiating in (8.5) under the integral sign, it is easy to check that $f_\Omega \in C^{m,\nu}(\Omega)$, respectively, $f_\Omega \in C_\Omega^{m,\nu}(\Omega)$. Further, define the operators T_Ω and S_Ω:

$$(T_\Omega u)(x) = \int_\Omega \mathcal{K}(x,y,u(y))\,dy, \qquad x \in \Omega, \quad S_\Omega u = T_\Omega u + f_\Omega.$$

Using (8.2) and (8.3) with $k = |\alpha| = |\beta| = 0$, it is easy to check that, for any $u, v \in L^\infty(\Omega)$ with $\|u\|_{0,\Omega} := \sup_{x \in \Omega} |u(x)| \le d$, $\|v\|_{0,\Omega} \le d$, we have

$$\begin{aligned} &\|T_\Omega u\|_{0,\Omega} \le b_1(d)\varepsilon_{\delta_0}, \\ &\|T_\Omega u - T_\Omega v\|_{0,\Omega} \le b_2(d,d)\varepsilon_{\delta_0}\|u-v\|_{0,\Omega}, \quad \varepsilon_{\delta_0} \to 0 \text{ as } \delta_0 \to 0. \end{aligned} \tag{8.6}$$

A consequence is that, for sufficiently small $\delta_0 > 0$, S_Ω maps the ball

$$\mathcal{B}_{0,\Omega,d} = \{u \in L^\infty(\Omega) : \|u\|_{0,\Omega} \le d\}, \qquad d > \|f_\Omega\|_{0,\Omega},$$

into itself and is contractive on it:

$$\|S_\Omega u - S_\Omega v\|_{0,\Omega} \le q\|u-v\|_{0,\Omega}, \quad q < 1, \qquad \forall u,v \in \mathcal{B}_{0,\Omega,d}. \tag{8.7}$$

Due to the Banach fixed point theorem, S_Ω has a unique fixed point in $\mathcal{B}_{0,\Omega,d}$; we know this fixed point – it is u_0, the solution of (8.1) under consideration. A more serious consequence of (8.2) and (8.3) is that, for sufficiently small $\delta_0 > 0$, the operator S_Ω maps a closed set $\mathcal{B}_{m,\nu,\Omega,d,d'}$ of $C^{m,\nu}(\Omega)$ into itself and is contractive on it with respect to a norm $\|u\|'_{m,\nu,\Omega}$ which is equivalent to the usual norm $\|u\|_{m,\nu,\Omega}$ of $C^{m,\nu}(\Omega)$:

$$\|S_\Omega u - S_\Omega v\|'_{m,\nu,\Omega} \le q\|u-v\|'_{m,\nu,\Omega}, \qquad q<1, \qquad \forall u,v\in \mathcal{B}_{m,\nu,\Omega,d,d'}. \tag{8.8}$$

The details follow soon; here we note that $\mathcal{B}_{m,\nu,\Omega,d,d'} \subset \mathcal{B}_{o,\Omega,d}$. Using again the Banach fixed point theorem we see that equation (8.4) is uniquely solvable in $\mathcal{B}_{m,\nu,\Omega,d,d'}$. The solution coincides with the unique solution u_o of equation (8.4) (equation (8.1)) in $\mathcal{B}_{o,\Omega,d}$. In other words, the restriction to Ω of u_o, the solution to (8.1) under consideration, belongs to $\mathcal{B}_{m,\nu,\Omega,d,d'} \subset C^{m,\nu}(\Omega)$. Repeating the argument from Section 3.5 we conclude that $u_o \in C^{m,\nu}(G)$.

A similar proof shows that, under conditions of Theorem 8.2, $u_o \in C_a^{m,\nu}(G)$.

To make the proof complete we have to establish the contractivity property (8.8). We present the outlines of the proof restricting ourselves to the case $m=2$. For any $u\in C^{2,\nu}(G)$ we have (cf. (3.11))

$$\frac{\partial}{\partial x_i}\int_\Omega \mathcal{K}(x,y,u(y))dy = \int_\Omega \Big(\frac{\partial}{\partial x_i}+\frac{\partial}{\partial y_i}\Big)(\mathcal{K}(x,y,u(y)))dy + \int_{\partial\Omega}\mathcal{K}(x,y,u(y))\,\omega_i(y)dS_y$$

$$= \int_\Omega \mathcal{K}_i(x,y,u(y))dy + \int_\Omega \mathcal{K}_o(x,y,u(y))\,\frac{\partial u(y)}{\partial y_i}dy + \int_{\partial\Omega}\mathcal{K}(x,y,u(y))\,\omega_i(y)dS_y, \tag{8.9}$$

$$\frac{\partial}{\partial x_j}\frac{\partial}{\partial x_i}\int_\Omega \mathcal{K}(x,y,u(y))dy = \int_\Omega \mathcal{K}_{ij}(x,y,u(y))dy + \int_\Omega \mathcal{K}_{io}(x,y,u(y))\,\frac{\partial u(y)}{\partial y_j}dy$$

$$+\int_\Omega \mathcal{K}_{oj}(x,y,u(y))\,\frac{\partial u(y)}{\partial y_i}dy + \int_\Omega \mathcal{K}_{oo}(x,y,u(y))\frac{\partial u(y)}{\partial y_i}\frac{\partial u(y)}{\partial y_j}dy + \int_\Omega \mathcal{K}_o(x,y,u(y))\,\frac{\partial^2 u(y)}{\partial y_i\partial y_j}dy$$

$$+\int_{\partial\Omega}\mathcal{K}_i(x,y,u(y))\omega_j(y)dS_y + \int_{\partial\Omega}\mathcal{K}_o(x,y,u(y)\frac{\partial u(y)}{\partial y_i}\omega_j(y)dS_y + \int_{\partial\Omega}\frac{\partial}{\partial x_j}\mathcal{K}(x,y,u(y))\omega_i(y)dS_y \tag{8.10}$$

where $\omega(y)=(\omega_1(y),\dots,\omega_n(y))$ with $\omega_i(y)=\dfrac{\bar{x}_i-y_i}{\delta}$ is the unit inner normal to $\partial\Omega = S(\bar{x},\delta)$ at $y\in\partial\Omega$

$$\mathcal{K}_i(x,y,u) = \Big(\frac{\partial}{\partial x_i}+\frac{\partial}{\partial y_i}\Big)\mathcal{K}(x,y,u), \qquad \mathcal{K}_o(x,y,u) = \frac{\partial}{\partial u}\mathcal{K}(x,y,u),$$

$$\mathcal{K}_{ij}(x,y,u) = \Big(\frac{\partial}{\partial x_j}+\frac{\partial}{\partial y_j}\Big)\Big(\frac{\partial}{\partial x_i}+\frac{\partial}{\partial y_i}\Big)\mathcal{K}(x,y,u), \qquad \mathcal{K}_{oo}(x,y,u) = \frac{\partial^2}{\partial u^2}\mathcal{K}(x,y,u),$$

$$\mathcal{K}_{oi}(x,y,u) = \mathcal{K}_{io}(x,y,u) = \Big(\frac{\partial}{\partial x_i}+\frac{\partial}{\partial y_i}\Big)\frac{\partial}{\partial u}\mathcal{K}(x,y,u), \qquad i,j,=1,\dots,n,$$

are weakly singular kernels (see (8.2), (8.3)). Note that the singularity of the term $\frac{\partial u}{\partial y_i}\frac{\partial u}{\partial y_j}$ is milder than the singularity allowed for $\frac{\partial^2 u}{\partial y_i \partial y_j}$ by the definition of the space $C^{2,\nu}(\Omega)$:

$$\left|\frac{\partial u(y)}{\partial y_i}\frac{\partial u(y)}{\partial y_j}\right| \le \text{const}\left\{\begin{array}{ll} 1 & , \ \nu < n-1 \\ (1+|\log\rho_\Omega(y)|)^2, & \nu = n-1 \\ \rho_\Omega(y)^{2(n-\nu)-2} & , \ \nu > n-1 \end{array}\right\},$$

$$\left|\frac{\partial^2 u(y)}{\partial y_i \partial y_j}\right| \le \text{const}\left\{\begin{array}{ll} 1 & , \ \nu < n-2 \\ 1+|\log\rho_\Omega(y)|, & \nu = n-2 \\ \rho_\Omega(y)^{n-\nu-2} & , \ \nu > n-2 \end{array}\right\}.$$

Recalling that $\Omega = B(\bar{x},\delta)$, $\delta = \min\{\delta_0, \rho(\bar{x})/2\}$, we estimate the terms of (8.9) and (8.10) by means of (8.2) and (8.3). We find that, for any $u,v \in C^{2,\nu}(G)$ with $\|u\|_{0,\Omega} \le d$, $\|v\|_{0,\Omega} \le d$, we have

$$w_{|\alpha|-(n-\nu),\Omega}(x)\,|D^\alpha(T_\Omega u)(x)| \le b_1(d)\,(c + \varepsilon_{\delta_0}\|u\|_{|\alpha|,\nu,\Omega}),$$

$$w_{|\alpha|-(n-\nu),\Omega}(x)|D^\alpha(T_\Omega u - T_\Omega v)(x)| \le c b_2(d,d)\,\|u-v\|_{0,\Omega} + b_1(d)\varepsilon_{\delta_0}\|u-v\|_{|\alpha|,\nu,\Omega}, \tag{8.11}$$

$$x \in \Omega, \quad 1 \le |\alpha| \le 2, \quad \varepsilon_{\delta_0} \to 0 \ \text{ as } \ \delta_0 \to 0,$$

where the constant c is independent of u, v and d. These inequalities may be extended to any $u,v \in C^{2,\nu}(\Omega)$ with $\|u\|_{0,\Omega} \le d$, $\|v\|_{0,\Omega} \le d$. Introduce a new norm $\|\cdot\|'_{2,\nu,\Omega}$ in $C^{2,\nu}(\Omega)$:

$$\begin{aligned} \|u\|'_{2,\nu,\Omega} &= M\,\|u\|_{0,\Omega} + \|u\|_{2,\nu,\Omega} \\ &= (M+1)\,\|u\|_{0,\Omega} + \sum_{1\le|\alpha|\le 2} \sup_{x\in\Omega} \left(w_{|\alpha|-(n-\nu),\Omega}(x)\,|D^\alpha u(x)|\right) \end{aligned}$$

where $M > 2c\max\{b_1(d), b_2(d,d)\}$. It is clear that the norms $\|\cdot\|_{2,\nu,\Omega}$ and $\|\cdot\|'_{2,\nu,\Omega}$ are equivalent. Introduce the set

$$\mathcal{B}_{2,\nu,\Omega,d,d'} = \{u \in C^{2,\nu}(\Omega) \colon \|u\|'_{2,\nu,\Omega} \le d', \ \|u\|_{0,\Omega} \le d\}, \qquad d' > \|f_\Omega\|'_{m,\nu,\Omega}.$$

It is clear that $\mathcal{B}_{2,\nu,\Omega,d,d'} \subset \mathcal{B}_{0,\Omega,d}$ and $\mathcal{B}_{2,\nu,\Omega,d,d'}$ is closed in $C^{2,\nu}(\Omega)$. A consequence of (8.6) and (8.11) is that, for sufficiently small δ_0, the operator S_Ω maps $\mathcal{B}_{2,\nu,\Omega,d,d'}$ into itself and satisfies (8.8) with $q = 1/2$.

The proof of Theorems 8.1 and 8.2 is finished.

8.2. Piecewise constant collocation and related methods. Let $G_{j,h}$, $j=1,\dots,l_h$, be an approximate partition of G described in Section 5.2 (i). We apply the piecewise constant collocation method

$$u_{i,h} = \sum_{j=1}^{l_h} \int_{G_{j,h}} \mathcal{K}(\xi_{i,h}, y, u_{j,h})dy + f(\xi_{i,h}), \quad i=1,\dots,l_h, \tag{8.12}$$

and the cubature formula method

$$u_{i,h} = \sum_{\substack{j=1 \\ \operatorname{dist}(\xi_{i,h}, \operatorname{co} G_{j,h}) \ge h}}^{l_h} \mathcal{K}(\xi_{i,h}, \xi_{j,h}, u_{j,h}) \operatorname{meas} G_{j,h} + f(\xi_{i,h}), \quad i=1,\dots,l_h, \tag{8.13}$$

to solve the nonlinear integral equation (8.1).

Theorem 8.3. Assume that the following conditions are fulfilled:

1. $G \subset \mathbb{R}^n$ is open and bounded, ∂G satisfies condition (PS) (see Section 5.1);
2. partition of G and collocation points satisfy (5.4) - (5.8);
3. kernel $\mathcal{K}(x,y,u)$ satisfies (8.2) and (8.3) with $m=2$;
4. $f \in C^{2,\nu}(G)$ satisfies (5.16) with $\mu = \nu$;
5. integral equation (8.1) has a solution $u_0 \in L^\infty(G)$ and the linearized integral equation

$$v(x) = \int_G K_0(x,y)v(y)dy, \quad K_0(x,y) = \left[\partial\mathcal{K}(x,y,u)/\partial u\right]_{u=u_0(x)}, \tag{8.14}$$

has in $L^\infty(G)$ only the trivial solution $v=0$.

Then there exist $h_0>0$ and $\delta_0>0$ such that system (8.12), as well system (8.13), has a unique solution $(u_{i,h})$ satisfying

$$\max_{1 \le i \le l_h} |u_{i,h} - u_0(\xi_{i,h})| \le \delta_0.$$

The following error estimates hold: for method (8.12),

$$\max_{1 \le i \le l_h} |u_{i,h} - u_0(\xi_{i,h})| \le \text{const}\, \varepsilon_{\nu,h}^2; \tag{8.15}$$

for method (8.13),

$$\max_{1 \le i \le l_h} |u_{i,h} - u_0(\xi_{i,h})| \le \text{const}\, \varepsilon'_{\nu,h} \tag{8.16}$$

with $\varepsilon_{\nu,h}$, $\varepsilon'_{\nu,h}$ defined in (5.18) and (5.23).

Proof. The proof can be constructed on the basis of Theorem 4.3. We consider (8.1) as equation $u = Tu + f$ in the space $E = BC(G)$ and system (8.12) as equation $u_h = T_h u_h + p_h f$ in the space $E_h = C(\Xi_h)$, compare with Section 5.5. We have assumed that (8.1) has a solution $u_0 \in L^\infty(G)$ but due to Theorem 8.1, $u_0 \in C^{2,\nu}(G) \subset BC(G)$. It is clear that the operators $T: E \to E$ and $T_h: E_h \to E_h$ are Frechet differentiable,

$$(T'(u_o)v)(x) = \int_G \frac{\partial \mathcal{K}(x,y,u_o(y))}{\partial u} v(y)\,dy,$$

$$(T'_h(u_h)v_h)(\xi_{i,h}) = \sum_{j=1}^{l_h} \int_{G_{j,h}} \frac{\partial \mathcal{K}(\xi_{i,h},y,u_{j,h})}{\partial u}\,dy\, v_{j,h},$$

$$\|T'_h(u_h) - T'_h(p_h u_o)\| \le \varepsilon \quad \text{whenever } \|u_h - p_h u_o\| \le \delta_\varepsilon .$$

We have (cf. (5.37)

$$\max_{1\le i\le l_h} \sup_{y\in G'_{j,h}} |u_o(\xi_{j,h}) - u_o(y)| \le \text{const}\, \varepsilon_{\nu,h}, \tag{8.17}$$

and using the technique of Section 5.7 we can estimate

$$\|T_h p_h u_o - p_h T u_o\| = \max_{1\le i\le l_h} \Big| \sum_{j=1}^{l_h} \int_{G_{j,h}} \mathcal{K}(\xi_{i,h}, y, u_o(\xi_{j,h}))\,dy - \int_G \mathcal{K}(\xi_{i,h},y,u_o(y))\,dy \Big|$$

$$\le \max_{1\le i\le l_h} \sum_{j=1}^{l_h} \Big| \int_{G'_{j,h}} \big[\mathcal{K}(\xi_{i,h},y,u_o(\xi_{j,h})) - \mathcal{K}(\xi_{i,h},y,u_o(y))\big]dy\Big| + c\varepsilon^2_{\nu,h} \tag{8.18}$$

$$\le \max_{1\le i\le l_h} \sum_{j=1}^{l_h} \Big| \int_{G'_{j,h}} \frac{\partial \mathcal{K}(\xi_{i,h},y,u_o(y))}{\partial u} [u_o(\xi_{j,h}) - u_o(y)]\,dy\Big| + c'\varepsilon^2_{\nu,h} \le c''\varepsilon^2_{\nu,h}.$$

Further, according to Lemma 5.2, $S_h \dashrightarrow T'(u_o)$ compactly where $S_h \in \mathcal{L}(E_h,E_h)$ is defined as the PCCM discretization of $T'(u_o)$,

$$(S_h v_h)(\xi_{i,h}) = \sum_{j=1}^{l_h} \int_{G_{j,h}} \frac{\partial \mathcal{K}(\xi_{i,h},y,u_o(y))}{\partial u}\,dy\, v_{j,h}.$$

Comparing this with

$$(T'_h(p_h u_o)(\xi_{i,h}) = \sum_{j=1}^{l_h} \int_{G_{j,h}} \frac{\partial \mathcal{K}(\xi_{i,h},y,u_o(\xi_{j,h}))}{\partial u}\,dy\, v_{j,h}$$

we conclude that $\|T'_h(p_h u_o) - S_h\| \le c\varepsilon_{\nu,h} \to 0$ as $h \to 0$ (see (8.17)) and $T'_h(p_h u_o) \dashrightarrow T'(u_o)$ compactly. Thus, for equation (8.1) and system (8.12), all conditions of Theorem 4.3 are fulfilled. Error estimate (4.19) immediately provides (8.15).

We treat system (8.13) also as an equation $u_h = \tilde{T}_h u_h + p_h f$ in the space $E_h = C(\Xi_h)$. The check of the conditions of Theorem 4.3 contains no new ideas. Error estimate (4.19) now yields (see (8.18))

$$\|u_h - p_h u_o\| \le c_2 \|\tilde{T}_h p_h u_o - p_h T u_o\| \le c_2(\|\tilde{T}_h p_h u_o - T_h p_h u_o\| + \|T_h p_h u_o - p_h T u_o\|)$$

$$\le c_2 \|\tilde{T}_h p_h u_o - T_h p_h u_o\| + c\,\varepsilon^2_{\nu,h}.$$

To obtain (8.16), it remains to show that $\|\tilde{T}_h p_h u_o - T_h p_h u_o\| \le c\,\varepsilon'_{\nu,h}$. This can be done with the help of the technique of Section 5.10.

The proof of Theorem 8.3 is finished.

8.3. Higher order methods. Now we consider the case where G is parallelepiped (7.1). We look for an approximate solution to integral equation (8.1) as a piecewise polynomial function $u_N \in E_N$ of degree m-1 with respect to any of arguments $x_1, \dots, x_n$ corresponding to grid (7.3). More precise definitions of E_N, the interpolation projector P_N and the collocation points see in Section 7.2. We determine $u_N \in E_N$ by the collocation method which we present here only in the operator equation form:

$$u_N = P_N T u_N + P_N f. \tag{8.19}$$

<u>Theorem</u> 8.4. Let the following conditions be fulfilled:

1. $G \subset \mathbb{R}^n$ is a parallelepided (see (7.1));
2. graded grid (7.3) and collocation points (7.7) are used;
3. kernel $\mathcal{K}(x,y,u)$ satisfies (8.2), (8.3);
4. $f \in C^{m,\nu}_{\square}(G)$ (see Section 7.1);
5. integral equation (8.1) has a solution $u_o \in L^\infty(G)$ whereby the linearized equation (8.14) has in $L^\infty(G)$ only the trivial solution $v = 0$.

Then there exist $N_k^o > 0$ $(k = 1, \dots, n)$ and $\delta_o > 0$ such that, for $N_k \ge N_k^o$ $(k = 1, \dots, n)$, the collocation method (8.19) defines a unique approximation $u_N \in E_N$ to u_o satisfying $\|u_N - u_o\|_{L^\infty(G)} \le \delta_o$. The following error estimates hold:

$$\max_{y \in \overline{G}} |u_N(x) - u_o(x)| \le c h^m \quad \text{for} \quad \left\{ \begin{array}{lll} r > m/(n-\nu) & \text{if} & n-\nu < m \\ r > 1 & \text{if} & n-\nu = m \\ r \ge 1 & \text{if} & n-\nu > m \end{array} \right\}; \tag{8.20}$$

$$\varepsilon_N \le c h^m \quad \text{for} \quad \left\{ \begin{array}{lll} r > m/[2(n-\nu)] & \text{if} & n-\nu \le 1 \\ r > m/(n-\nu+1) & \text{if} & 1 < n-\nu \le m-1 \\ r \ge 1 & \text{if} & n-\nu > m-1 \end{array} \right\} \tag{8.21}$$

where

$$\varepsilon_N = \max_{i_k = 1, \dots, m,\ j_k = 0, \dots, 2N_k - 1,\ k = 1, \dots, n} |u_N(x) - u_o(x)|_{x = (\xi_1^{j_1, i_1}, \dots, \xi_n^{j_n, i_n})}$$

is the maximal error of u_N at collocation points.

Proof. Let us treat (8.1) and (8.19) as equations (4.17) and (4.18), respectively. Putting $E=E_h=L^\infty(G)$, $p_h=I$, it is easy to check that the operators $T:L^\infty(G)\to L^\infty(G)$ and $P_NT: L^\infty(G)\to L^\infty(G)$ satisfy the conditions (i)-(iv) of Theorem 4.3. Error estimate (4.19) provides

$$\|u_N-u_0\|_{L^\infty(G)} \le c_2\|u_0-P_Nu_0\|_{L^\infty(G)}.$$

It follows from conditions 3 and 4 that (see Theorem 8.2 and Section 7.1) that $u_0\in C_\square^{m,\nu}(G)$. Using Lemma 7.2 we see that (8.20) is true.

To prove (8.21) we choose other spaces and connection operators: $E=BC(G)$, $E_h=E_N$ (the space of piecewise polynomials defined in Section 7.2 and equipped with the supremum-norm), $p_h=P_N$. One can check again that the operators $T:BC(G)\to BC(G)$ and $P_NT:E_N\to E_N$ satisfy the conditions (i)-(iv) of Theorem 4.3. Now estimate (4.19) yields

$$\|u_N-P_Nu_0\|_{L^\infty(G)} \le c_2\|P_NTP_Nu_0-P_NTu_0\|_{L^\infty(G)} \le c\|TP_Nu_0-Tu_0\|_{L^\infty(G)},$$

or (see (7.33) and (8.2), (8.3))

$$\varepsilon_N \le c \sup_{y\in\overline{G}} \left|\int_G\left[\mathcal{K}(x,y,(P_Nu_0)(y)) - \mathcal{K}(x,y,u_0(y))\right]dy\right|$$

$$\le c \sup_{y\in\overline{G}} \left|\int_G \frac{\partial\mathcal{K}(x,y,u_0(y))}{\partial u}[(P_Nu_0)(y)-u_0(y)]dy\right| + c'\|u_0-P_Nu_0\|^2_{L^\infty(G)} \quad (8.22)$$

$$\le c''\int_G \begin{Bmatrix} 1 & , & \nu<0 \\ 1+|\log|x-y||, & & \nu=0 \\ |x-y|^{-\nu} & , & \nu>0 \end{Bmatrix} |u_0(y)-(P_Nu_0)(y)|dy + c'\|u_0-P_Nu_0\|^2_{L^\infty(G)}.$$

Using Lemmas 7.2 and 7.3 we obtain (8.21).

The proof of Theorem 8.4 is finished.

Inequality (8.22) can be used as the key to extend Theorem 7.4, too, to the case of nonlinear integral equation (8.1). Compared with Section 7.6, we have a complication: together with $D^\alpha u_0(y)$, the derivatives $D^\alpha_y K_0(x,y)$, $K_0(x,y) = \partial\mathcal{K}(x,y,u_0(y))/\partial u$, have singularities near ∂G.

8.4. Exercises. There is much work to fill the details of the proofs of Theorems 8.1-8.4. Further, one can try to extend other results of Chapters 5-7 to nonlinear integral equations. Actually, here the exercises grow up to a real research since there are only short results in the literature concerning the solution of nonlinear integral equations with singularities. For instance, one could construct and examine two grid iteration methods (cf. Section 5.12) on the basis of the Newton method.

BIBLIOGRAPHICAL COMMENTS

General. Let us point out some books on integral equations and specially on their approximate solution: Mikhlin (1964), Anselone (1971), Atkinson (1976), Baker (1977), Fenyö and Stolle (1984), Reinhardt (1985), Hackbusch (1989), Kress (1989).

To Chapter 1. The integral equation formulations of the interior-exterior problem of Section 1.1 were considered by Vainikko (1990 b). The Peierls integral equations are broadly known in literature on the theory of radiation transfer, see e.g. Marchuk and Lebedev (1984) or Smelov (1978); in Section 1.6, integral equation (1.15) corresponding to general phase function of scattering was obtained following a similar argument. About methods to solve radiation transfer problems see also Germogenova (1986), Knyazikhin and Marshak (1987), Vainikko and Marshak (1978), Vainikko (1990 d).

To Chapter 2. In the case where G is a regular region such that $G^* = \overline{G}$, assertions like Lemmas 2.1– 2.3 are well known. Our treatment was caused by a desire to embrace the case of piecewise smooth K(x,y) and f(x) on $\overline{G} \times \overline{G}$ and $\overline{G}$, respectively. Formally this was achieved introducing and using the inner distance d_G instead of the Euklidean distance. The proof of Lemma 2.2 is somewhat unusual; traditional proofs do not suit here.

To Chapter 3. The smoothness of the solutions to an one dimensional weakly singular integral equations was examined by Richter (1976), Pedas (1979), Schneider(1979), Vainikko and Pedas (1981), Graham(1982 a), Vainikko, Pedas and Uba (1984). In the last work the one dimensional counterparts of Theorems 3.1 and 3.3 are given. Uba (1988 a) and R.Kangro (1990) considered the situation where the singularities of the derivatives of the solution may occur in the inner points of the interval of integration. This takes place e.g. if the singularity of the kernel is concentrated on a curve $y = \varphi(x)$ instead of the diagonal $x = y$.

There are only few works concerning the multidimensional case. The main landmarks have been put down by Pitkäranta (1979, 1980). Theorems 3.1 and 3.2 follow Vainikko (1991 a) but are supplied with new, probably more elementary proofs. We already refered to the papers of U. Kangro (1990 a, 1992) and partly formulated her results concerning the logarithmic kernel (Theorem 3.4). We also refer to Günter (1967) for properties of the Newton potential.

To Chapter 4. In a non-formalized manner, the discrete convergence of a sequence or a family (u_h), $u_h \in E_h$, has been used for a long time. Stummel (1970) gave a successful formalization of this notion; see also Vainikko (1976 a,b, 1977 b), Karma (1992 a,b).

Under other name, the concept of compact convergence of operators has been introduced by Sobolev (1956). Independently Anselone and Moore (1964), Anselone (1971) discovered and opened the utility of the concept ("collectively compact approximation"). In these works the concept of compact convergence of operators was used in the framework of usual norm convergence of elements ($E_h = E$, $p_h = I$). Together with some kinds of discrete convergence of elements the concept was developed by Vainikko (1968, 1969, 1970) and to full extent by Stummel (1970). Nowadays there is an extended literature on works where the compact convergence as well a related concept of regular convergence of operators is examined and/or used as a base, see e.g. Anselone and Ansorge (1981, 1987), Anselone and Treuden (1985), Gähler (1985), Grigorieff (1973, 1975), Heinrich (1983), Karma (1971 a,b, 1983, 1989, 1990, 1992 a,b), Stummel (1970, 1971, 1972, 1973, 1975, 1976), Vainikko (1974 a,b, 1976 a,b, 1977 a,b, 1978, 1981 a,b, 1983), Vainikko and Karma (1974 a,b 1988), Vainikko and Piskarev (1977), Wolf (1974, 1979).

To Chapter 5. Theorem 5.1 is based on the papers of Vainikko (1990 a, 1991 b) and Vainikko and Pedas (1989, 1990). Earlier Graham (1981) examined PCCM for weakly singular integral equations on a rectangle but his error estimates are of non-optimal order. Schneider (1981) and Graham and Schneider (1985) have developed a related product integration method.

The quadrature and cubature formula methods for integral equations have served as a test field of different abstract theories, see e.g. Krasnoselski and al (1972), Vainikko (1968, 1970, 1971, 1976 a,b), Anselone (1971); concerning one dimensional weakly singular integral equations see Anselone (1981), Vainikko, Pedas and Uba (1984), Pedas (1987, 1992) where the substraction on the singularity (the trick of Kantorovich - Krylov) is exploited.

Two grid iteration methods for integral equations originate from works of Brakhage (1960) and Atkinson (1973). An abstract setting with compact convergence of operators was introduced and examined by Daugavet (1980). Uba (1983), Vainikko, Pedas and Uba (1984) used these methods for one dimensional weakly singular integral equations giving a convergence analysis in an abstract setting but without an analysis of the amount of arithmetical work. A systematical analysis of the convergence rates and the amount of arithmetical work for two and multi-grid iteration methods was given by Hackbusch (1989). His abstract results can be applied to methods (5.41) – (5.44), too, but this way is more complicated and the results will be somewhat weaker compared with the direct analysis presented in Sections 5.12 - 5.14. We followed here Vainikko (1992 a); see also Vainikko (1990 b) where the case $\nu = n-2$ was analysed.

To Chapter 6. This Chapter is based on the paper of Vainikko (1992 b) inspirated by Georg (1991) and Allgower, Georg and Widmann (1992). Joe (1985 a), Atkinson and Flores (1992) examined fully discretized collocation methods, Joe (1987), Atkinson and Bogomolny (1987), Atkinson and Potra (1988, 1989) – fully discretized Galerkin methods, but the amount of arithmetical work has not been discussed in those works. Johnson and Scott (1989) constructed discretizations of an order minimal amount of arithmetical work for boundary integral equations. We refer also to Uba (1992) where the computation of weakly singular integrals is discussed.

To Chapter 7. Rice (1969) seems to be the first who introduced and used the graded grids of type (7.3). In the one dimensional case ($n=1$) Theorems 7.1 - 7.3 were proved by Vainikko and Uba (1981), in the multidimensional case by Vainikko (1988); see also Uba (1989) , U. Kangro (1990 b).

The superconvergence of the collocation method has been examined by Prenter (1973), Chatelin and Lebbar (1981), Chatelin (1983), Atkinson, Graham and Sloan (1983), Graham and Sloan (1985), Joe (1985 b), Brunner (1985), Sloan (1990); see also the works of Atkinson cited above. In a part of those works the superconvergence of Galerkin method is disscussed, too; to the last problem, the works of Richter (1978), Graham (1982b), Spence and Thomas (1983), Sloan and Thomee (1985) are devoted. Nevertheless, these works do not cover Theorem 7.4 since mostly they concern the one dimensional case with a smooth kernel without singularities or, at least, without singularities of the exact solution.

To Chapter 8. Theorems 8.1 and 8.2 were formulated by Vainikko (1990 c) but the proof sketch was more superficial than in the Chapter text. The results concerning the collocation method for nonlinear equations (Theorems 8.3 and 8.4) are presented here for the first time. Let us cite again the works of Atkinson and Potra (1988), Atkinson and Flores (1992).

About boundary integral equations. We have almost avoided the references on works about boundary integral equations (BIE). Let us now present some. In the books of Mikhlin (1970) and Schatz, Thomée and Wendland (1990) one can find the BIE reformulations of standard boundary value problems; in the survey by Atkinson (1990) one finds the algorithms for the numerical solving of BIE, too. We refer also to Kozlov, Lifanov and Mikhailov (1991) where some non-standard BIEs are proposed. The collocation and Galerkin methods for BIE are examined by Hsiao and Wendland (1981), Wendland (1980, 1983), Arnold and Wendland (1983), Costabel and Stephan (1987), Saranen and Wendland (1985), Saranen (1989); see also Mokin (1987, 1988). The knowledges about the singularities of the solution to BIE seem to be rather incomlete if the boundary is non-regular; some descriptions have been given by Costabel (1983), Maz'ya (1988), Schmitz (1989), Elschner (1991).

About singular integral equations. There is an abundant literature on the (strongly) singular integral equations. We confine ourselves referring to the books of Mikhlin (1965), Prössdorf and Silbermann (1977, 1991), Gohberg and Krupnik (1979), Belotserkovsky and Lifanov (1985), Mikhlin and Prössdorf (1986); see also Gabdulhaev (1980), Berthold and Junghanns (1992).

OPEN PROBLEMS

1. Theorem 3.3 and the examples of Section 3.10 show that in general the derivatives of a solution to integral equation (3.1) have singularities along whole boundary ∂G. On the other hand, the kernel $K(x,y) = a(x,y)\log|x-y|$, $n=2$, is exceptional — the singularities of the derivatives of a solution may occur only at singular points (e.g. corners) of ∂G, see Theorem 3.4. A similar exceptional kernel seems to be $K(x,y) = a(x,y)|x-y|^{-1}$, $n=3$. Note that $c_2\log|x-y|$ $(n=2)$ and $c_3|x-y|^{-1}$ $(n=3)$ are fundametal solutions to Laplace equation. Is this "exceptional" smoothness phenominon related to fundamental solutions of elliptic differential equations only? Otherwise, can one give general algebraic or analytical conditions on the kernel $K(x,y)$ for the "exceptional" smoothness of the solutions to (3.1)?

2. Theorem 3.3 presents the main singular part of a derivative of a solution to integral equation (3.1). It were interesting to find a decomposition of a solution detaching all terms with singular derivatives of a given order m. For $n=1$, such a decomposition is derived by Graham (1982a). For $n\geq 2$, Pitkäranta (1980) gave a decomposition in the case of kernel $K(x,y)=a(x,y)|x-y|^{-\nu}$, $\nu<n$, but it were interesting to know more about the coefficients of the decomposition. Can one find a decomposition such that both Theorems 3.1 and 3.2 are consequences from it?

3. Let $a^j(x)=(a_1^j(x), \dots , a_n^j(x))$, $j=1,2, \dots$, be tangential vector fields to $\partial\Omega$ of the class $C^m(\overline{G})$. It were interesting to characterize the behaviour of the "mixed" tangential derivatives $L_{a^{j_k}} \dots L_{a^{j_1}}u(x)$ of a solution to integral equation (3.1). Theorem 3.2 concerns the case where all $a^j(x)$, $j=1,2,\dots$, coincide.

4. There are some problems from praxis which can be formulated as an one dimensional integral equation

$$u(x) = \int_0^1 K(x,y)\,u(y)\,dy + f(x)$$

where the kernel $K(x,y)$ has a singularity along a curve $\Phi(x,y)=0$. It is little known about the smoothness of a solution of such equation. A more special case is examined by R. Kangro (1990). It were interesting to investigate the problem in the general setting formulated above and to generalize the results to multidimensional equations.

5. It were interesting to realize the ideas of Chapter 6 for the higher order collocation methods considered in Chapter 7. Also the result about the two grid iteration methods (Chapter 5) should be extended.

6. In this book we have restricted ourselves to collocation methods. Similar results can be developed for the Galerkin method. Let us point out an advantage of collocation method: one can use the benefits of the super-

convergence of the collocation solution $u_N(x)$ at the collocation points without any supplementary work. Perhaps this advantage is only seeming, and the Galerkin solution $v_N(x)$ has a similar property at some special points? In the literature, the superconvergence is established for

$$\tilde{v}_N(x) = \int_G K(x,y)\, v_N(y)\, dy + f(x), \qquad x \in G,$$

but the evaluation of the integral needs supplementary calculations.

7. Serious difficulties occur when one tries to extend the results of Chapter 7 to the case of a region G with a curved smooth or piecewise smooth boundary ∂G. A natural way to construct high order collocation or Galerkin schemes seems to be the use of isoparametric finite elements approximating the solution as well the boundary. About the isoparametric elements see e.g. Ciarlet (1978).

8. For one dimensional BIE $u = Tu + f$ on a smooth closed curve Γ, rather effective Galerkin method $u_N = P_N T u_N + P_N f$ can be constructed using orthoprojector P_N to N-dimensional wavelet subspace V_N of $L^2(\Gamma)$; about orthogonal wavelet bases see Daubechies (1988), Meyer (1990). The error $\|u_N - u\| \le c\,\|u - P_N u\|$ is of order $\mathcal{O}(N^{-m})$ provided that the exact solution u belongs to $H^m(\Gamma)$ and appropriate wavelets are used. The operator $P_N T$ can be applied to a function $u_N \in V_N$ with a given accuracy ε (e.g. $\varepsilon = N^{-m}$) in $\mathcal{O}(N \log N)$ arithmetical operations provided that the kernel of the BIE is smooth, too; see Beylkin, Coifman and Rokhlin (1991). This enables to solve the Galerkin equation $u_N = P_N T u_N + P_N f$ in $\mathcal{O}(N \log N)$ arithmetical operations using the analogues of the two grid methods considered in Section 5.12. It were interesting to examine the wavelet techniques for weakly singular integral equations on curves, surfaces as well domains $G \subset \mathbb{R}^n$.

9. For the radiation transfer integral equation (1.15), the methods considered in Chapters 5 and 7 yield only semidiscretizations — the argument $s \in S$ remains to be non-discretized. The convergence results should be revisited modifying the compact convergence techniques or using other ideas. It were also interesting to examine the full discretizations using with respect to $s \in S$ quadrature formulas or Galerkin approximations with sphere functions. This way were a counterbalance to the standard approaches where the discretizations are introduced directly into the integrodifferential equation (1.11), see e.g. Chandrasekhar (1950), Marchuk and Lebedev (1984), Smelov (1978), Germogenova (1986).

10. Turning once more to the radiation transfer problem, consider a physically interesting case where $\sigma : \overline{G} \to \mathbb{R}$ is piecewise smooth with breaks along some surfaces between subregions G_j, $G_j \subset G$. Strange enough, the smoothness properties of the optical distance

$$\Sigma(x,y) = |x-y| \int_0^1 \sigma(\lambda x + (1-\lambda)y)d\lambda, \qquad x,y \in G,$$

have not been seriously examined, and this problem is not so easy as it may seem at the first look. Having done this, it were interesting to know how behave the derivatives of the solution to the Peierls integral equation (1.17) as well to the general radiation transfer integral equation (1.15). Probably this information allows to improve the discretization schemes for those equations.

11. It were interesting to extend the results of the book to integral equation (3.1) on a unbounded region $G \subset \mathbb{R}^n$. Perhaps the following two cases are of greater interest: (i) the equation

$$u(x) = \int_G K(x,y)\, a(y)\, u(y)\, dy + f(x)$$

with a smooth $a \in L^1(G)$ and K satisfying (3.2); (ii) the equation of the Wiener-Hopf type,

$$u(x) = \int_G a(x,y)\, k(x-y) u(y)\, dy + f(x)$$

where $G = \{x \in \mathbb{R}^n:\ 0 < x_j < \infty,\ j = 1, \ldots, n\}$, a is smooth and bounded, $k \in L^1(\mathbb{R}^n)$ is smooth for $x \neq 0$ and possibly weakly singular ($|k(x)| \le b(1+|x|^{-\nu})$, $\nu < n$). The Wiener-Hopf equation corresponds to $a(x,y) \equiv 1$. About one dimensional Wiener-Hopf equation, see Gohberg and Feldman (1974); the collocation method is examined by Elschner (1989, 1990); see also Anselone and Sloan (1985), Sloan and Spence (1986), Chandler and Graham (1988a, b).

12. It were interesting to develop the results of Chapter 8 for more general nonlinear integral equation

$$f\Big(x, u(x), \int_G \mathcal{K}(x,y,u(y))dy\Big) = 0, \qquad x \in G.$$

13. The conjecture formulated in the Preface and concerning BIEs should be made more precise and proved. There stand also difficult problems concerning the smoothness of the solution to BIE on a non-smooth boundary ∂G.

REFERENCES

Akhmerov R.R. Kamenskij M.I., Potapov A.S., Rodkina A.E., Sadovskij B.N. (1992). *Measures of Noncompactness and Condensing Operators*. Birkhäuser.

Allgower E.L. Georg K., Widmann R. (1992). *Volume integrals for boundary elements methods*. J. Comput. Appl. Math. (to appear).

Anselone P.M. (1971). *Collectively Compact Approximation Theory*. Prentice-Hall: New Jersey.

Anselone P.M. (1981). *Singularity substraction in the numerical solution of integral equations*. J. Austral. Math. Soc. (Ser. B) 22, 408-418.

Anselone P.M., Ansorge R. (1981). *A unified framework for the discretization of nonlinear operator equations*. Numer. Funct. Anal. and Optim. 4, No.1, 61-99.

Anselone P.M., Ansorge R. (1987). *Discrete closure and asymptotic (quasi)-regularity in discretization algorithms*. IMA J. Numer. Anal. 7, 431-448.

Anselone P.M., Moore R.H. (1964). *Approximate solution of integral and operator equations*. J. Math. Anal and Appl. 9, No.2, 268-277.

Anselone P.M, Sloan I.H. (1985). *Integral equations on the half-line*. J. Integr. Equat., V.9 (Suppl.), 3-23.

Anselone P.M., Treuden M.L. (1985). *Regular operator approximation theory*. Pac. J. Math. 120, No.2, 257-268.

Arnold D.N., Wendland W.L. (1983). *On the asymptotic convergence of collocation methods*. Math. Comp. 41, 349-381.

Atkinson K.E. (1973). *Iterative variants of the Nyström method for the numerical solution of integral equations*. Numer. Math. 22, 17-31.

Atkinson K.E. (1976). *A Survey of Numerical Methods for the Solution of Fredholm Integral Equations of the Second Kind*. SIAM: Philadelphia, Pennsylvania.

Atkinson K.E. (1990). *A survey of boundary integral equations methods for the numerical solution of Laplace's equation in three dimensions*. In: Numerical Solution of Integral Equations, ed. M. Golberg, Plenum Press: New York.

Atkinson K.E., Bogomolny A. (1987). *The discrete Galerkin method for integral equations*. Math. Comp. 48, 595-616.

Atkinson K.E., Flores J. (1992). *The discrete collocation method for nonlinear integral equations*. IMA J. Num. Anal. (to appear)

Atkinson K.E., Graham I., Sloan I. (1983). *Piecewise continuous collocation for integral equations*. SIAM J. Numer. Anal., V.20, No.1, 172-186.

Atkinson K.E., Potra F.A. (1988). *The discrete Galerkin method for nonlinear integral equations*. J. Integr. Eq. and Appl. 1, 17-54.

Atkinson K.E., Potra F.A. (1989). *On the discrete Galerkin method for Fredholm integral equations of the second kind*. IMA J. Num. Anal. 9, 385-403.

Baker C.T.H. (1977). *The Numerical Treatment of Integral Equations*. Clarendon Press: Oxford.

Belotserkovsky S.M., Lifanov I.K. (1985). *Numerical Methods for Singular Equations and their Application in the Theory of Airfoils, Elasticity Theory and Electrodynamics*. Nauka: Moscow (in Russian).

Berthold D., Junghanns P. (1992). *On the pointwise convergence of the classical collocation methods for Cauchy-type singular integral equations*. SIAM J. Numer. Anal. (to appear).

Beylkin G., Coifman R., Rokhlin V. (1991). *Fast wavelet transforms and numerical analysis, I.* Comm. Pure Appl. Math., V.44, 141-193.

Bers L., John F., Schechter M. (1964). *Partial differential equations*. Interscience publ.: New York etc.

Brakhage H. (1960). *Über die numerische Behandlung von Integralgleichungen nach der Quadraturformelmethode*. Numer. Math. 2, 183-196.

Brunner H. (1985). *The numerical solution of weakly singular Volterra integral equations by collocation on graded meshes*. Math. Comp. 45, 417-437.

Case K.M., Zweifel P.M. (1967). *Linear Transport Theory*. Addison-Wesley Publ.: Reading, Massachusetts.

Chandler G.A., Graham I.G. (1988a). *Product integration-collocation methods for non-compact integral operator equations*. Math. Comp., V.50, 125-138.

Chandler G.A., Graham I.G. (1988b). *The convergence of Nyström methods for Wiener-Hopf equations*. Numer. Math., V.52, 345-364.

Chandrasekhar S. (1950). Radiative Transfer. Oxford.

Chatelin F. (1983). *Spectral Approximation of Linear Operators*. Academic Press: New York.

Chatelin F., Lebbar R. (1981). *The iterated projection solution for the Fredholm integral equations of the second kind*. J. Austral. Math. Soc. (Ser. B) 22, 439-451.

Ciarlet P.G. (1978). *The Finite Element Method for Elliptic Problems*. North-Holland: Amsterdam.

Costabel M. (1983). *Boundary integral operators on curved polygons*. Ann. Math. Pur. et Appl. 33, 305-326.

Costabel M., Stephan E. (1987). *On the convergence of collocation methods for boundary integral equations on polygons*. Math. Comp. 49, 461-478.

Crouzeix M., Descloux J. (1988). *A Bidimensional Electromagnetic Problem*. Rennes, Univ. de Rennes.

Daubechies I. (1988). *Orthogonal bases of compactly supported wavelets*. Comm. Pure Appl. Math., V.41, 909-996.

Daugavet I.K. (1980). *On the iterative solution of equations arising by compact approximation of operators*. U.S.S.R. Comp. Math. and Math. Phys. 20, No.4, 1046-1049 (in Russian).

Dieudonné J. (1960). *Foundations of Modern Analysis*. Academic Press: New York, London.

Elschner J. (1989). *On spline collocation for convolution equations*. Integr. Equat. and Oper. Theor., V.12, 486-510.

Elschner J. (1990). *On spline approximation for a class of non-compact integral equations*. Math. Nachr., V.146, 271-321.

Elschner J. (1991). *On the double layer potential operator over polyhedral domains: solvability in weighted Sobolev spaces and spline approximation*. In: Surveys in Analysis, Geometry and Math. Physics. Teubner.

Fenyö S., Stolle H.W. (1984). Theorie und Praxis der linearen Integralgleichungen. Bd.1-4. VEB Deutscher Verlag: Berlin.

Gabdulhaev B.G. (1980). *Finite Dimensional Approximations to Singular Integrals and Direct Methods for Singular Integral and Integro-differential Equations*. Itogi Nauki i Tehniki, Math. Anal., 18, Moscow.

Gähler S. (1985). *Discrete convergence*. Math. Res. 24, 127 - 136.

Georg K. (1991). *Approximation of integrals for boundary element methods*. SIAM J. Sci. Stat. Comput. 12, 443 - 453.

Germogenova T.A. (1986). *Local Properties of Solution to the Transport Equation*. Nauka: Moscow (in Russian).

Gohberg I.C., Feldman I.A., (1974). *Convolution Equations and Projection Methods for their Solution*. Amer. Math. Soc.: Providence.

Gohberg I., Krupnik N. (1979). *Einführung in die Theorie der eindimensionalen singulären Integralgleichungen*. Birkhäuser: Basel.

Graham I.G. (1981). *Collocation methods for two dimensional weakly singular integral equations*. J. Austral. Math. Soc. (Ser. B) 22, 456 - 473.

Graham I.G. (1982a). *Singularity expansions for the solutions of second kind Fredholm integral equations with weakly singular convolution kernels*. J. Integr. Equat. 4, No1, 1 - 30.

Graham I.G. (1982b). *Galerkin methods for second kind integral equations with singularities*. Math. Comput. 39, 519 - 533.

Graham I.G., Schneider C. (1985). *Product integration for weakly singular integral equations in $\mathbb{R}^n$*. In: Constructive Methods for the Practical Treatment of Integral Equations, ed. G. Hämmerlin and K.-H. Hoffmann. Birkhäuser Verlag.

Graham I.G., Sloan I.H. (1985). *Iterated Galerkin versus iterated collocation for integral equations of the second kind*. IMA J. Numer. Anal. 5, 355-369.

Grigorieff R.D. (1973). *Zur Theorie linearer approximationsregulärer Operatoren, I, II*. Math. Nachr. 55, 233 - 249, 251 - 263.

Grigorieff R.D. (1975). *Diskrete Approximation von Eigenwertproblemen, I - III*. Numer. Math. 24, 355 - 374, 415 - 433; 25, 79 - 97.

Günter N. (1967). *Pontential Theory*. Ungar: New York.

Hackbusch W. (1989). *Integralgleichungen*. Teubner: Stuttgart.

Heinrich S. (1983). *A problem in discretization theory related to the approximation property in Banach spaces*. Math. Nachr. 111, 147 - 152.

Hsiao G.C., Wendland W.L. (1981) . *The Aubin-Nitsche lemma for integral equations*. J. Integr. Eq. 3, 299 - 315.

Ivanov V.V. (1986). *Numerical Methods for Computers*. Naukova Dumka: Kiev (in Russian).

Joe S. (1985a). *Discrete collocation methods for second kind Fredholm integral equations*. SIAM J. Numer. Anal. 22, 1167 - 1177.

Joe S. (1985b). *Collocation methods using piecewise polynomials for second kind integral equations*. J. Comp. and Appl. Math. 12&13, 391 - 400.

Joe S. (1987). *Discrete Galerkin method for Fredholm integral equations of the second kind*. IMA J. Numer. Anal. 7, 149-164.

Johnson C., Scott L.R. (1989). *An analysis of quadrature errors in second-kind boundary integral equations*. SIAM J. Numer. Anal. 26, 1356 - 1382.

Kangro R. (1990). *On the smoothness of solutions to an integral equation with a kernel having a singularity on a curve*. Acta et comm. Univ. Tartuensis 913, 24 - 37.

Kangro U. (1990a). *The smoothness of the solution of a two-dimensional integral equation with logarithmic kernel*. Proc. Eston. Acad. Sci., phys., math. 39, 196 - 204 (in Russian).

Kangro U. (1990b). *Collocation method with bilinear splines for two-dimensional integral equation with logarithmic kernel*. Acta et comm. Univ. Tartuensis 913, 18 - 23 (in Russian).

Kangro U. (1992). *The smoothness of the solution to a two-dimensional integral equation with logarithmic kernel*. Z. Anal. und Anwend. (to appear).

Kantorovich L.V., Akilov G.P. (1982). *Functional Analyisis*. Pergamon Press.

Kantorovich L.V., Krylov V.I. (1962). *Approximative Methods of Higher Analysis*. Gos. Izd. Fiz.-Mat. Lit.: Moscow, Leningrad.

Karma O. (1971a). *Asymptotic error estimates of approximate characteristic values of holomorphic Fredholm operator functions*. U.S.S.R. Comp. Math. and Math. Phys. 11, No.3, 559-568 (in Russian).

Karma O. (1971b). *On the compact approximation of operator functions*. Acta et comm. Univ. Tartuensis 277, 194-204 (in Russian).

Karma O. (1983). *Approximation in eigenvalue problems with holomorphic dependence on parameter*. Acta et comm. Univ. Tartuensis 633, 19-28 (in Russian).

Karma O. (1989). *On the convergence of eigenvalues by approximation of the problem*. Acta et comm. Univ. Tartuensis 863, 40-56.

Karma O. (1990). *The convergence rate of approximate eigenvalues in projection-like methods*. Acta et comm. Univ. Tartuensis 913, 54-64.

Karma O. (1992a). *On the concept of discrete convergence in case of normed spaces*. Acta et comm. Univ. Tartuensis 937, 47-62.

Karma O. (1992b). *On the concept of discrete convergence*. Numer. Funct. Anal. and Optim. 13, No 1-2, 129-142.

Knyazikhin Yu. , Marshak A. (1987). *The Method of Discrete Ordinates for Solving the Transport Equation*. Valgus: Tallinn (in Russian).

Kozlov S.V., Lifanov I.K., Mikhailov A.A. (1991). *A new approach to mathematical modelling of flow of ideal fluid around bodies*. Sov. J. Numer. Anal. Math. Modelling, Vol.6, No.3, 209-222.

Krasnoselskii M.A., Vainikko G.M., Zabreiko P.P., Rutitskii Ya. B., Stetsenko V.Ya. (1972). *Approximate Solution of Operator Equations*. Wolters-Noordhoff: Groningen.

Kress R. (1989). *Linear Integral Equations*. Springer: Berlin, Heidelberg etc.

Lions J.-L., Magenes E. (1968). *Problèmes aux Limites non Homogènes et Applications*. Vol. 1. Dunod: Paris.

Marchuk G.I., Lebedev V.I. (1984). *Numerical Methods in the Neutron Transfer Theory*. Harwood Acad. Publ.: Chur.

Maz'ya V.G. (1988). *Boundary integral equations*. In: Sovrem. Problemy Mathem. Fundam. Napravlenia, Vol. 27, Viniti: Moscow, 131-228.

Meyer Y. (1990). *Ondellets*. Hermann: Paris.

Mikhlin S. G. (1964). *Integral Equations*. Pergamon Press: London.

Mikhlin S.G. (1965). *Multidimensional Singular Integrals and Integral Equations*. Pergamon Press: Oxford.

Mikhlin S. (1970). *Mathematical Physics: An Advanced Course*. North-Holland: Amsterdam.

Mikhlin S.G., Prössdorf S. (1986). *Singular Integral Operators*. Akademie-Verlag: Berlin.

Mokin Yu.I. (1987). *Numerical methods for integral equations of potential theory*. Differ. Uravn., V23, No.7, 1250-1262 (in Russian).

Mokin Yu. I. (1988). *Direct and iteration methods for integral equations of potential theory*. U.S.S.R. Comp. Math. Math. Phys., V.28, No.5, 669-682 (in Russian).

Pedas A. (1979). *On the smoothness of the solution of an integral equation with a weakly singular kernel*. Acta et comm. Univ. Tartuensis 492, 56-68 (in Russian).

Pedas A. (1987). *The numerical solution of weakly singular integral equation by quadrature method with trapezoidal formula*. Acta et comm. Univ. Tartuensis 762, 89-97 (in Russian).

Pedas A. (1992). *On the numerical solution of a weakly singular integral equation*. Acta et comm. Univ. Tartuensis 937, 15-26.

Pitkäranta J. (1979). *On the differential properties of solutions to Fredholm equations with weakly singular kernels*. J. Inst. Math. Phys. 24, 109-119.

Pitkäranta J. (1980). *Estimates for derivatives of solutions to weakly singular Fredholm integral equations*. SIAM J. Math. Anal. 11, 952-968.

Prenter P.M. (1973). *A collocation method for the numerical solution of integral equations*. SIAM J. Numer. Anal. 10, 570-581.

Prössdorf S., Silbermann B. (1977). *Projectionsverfahren und die näherungsweise Lösung singulären Gleichungen*. Treubner: Leipzig.

Prössdorf S., Silbermann B. (1991). *Numerical Analysis for Integral and Related Operators Equations*. Akademie-Verlag: Berlin.

Reinhardt H.-J. (1985). *Analysis of Appromation Methods for Differential and Integral Equations*. Springer: New York, Berlin, Heidelberg, Tokyo.

Rice J.R. (1969). *On the degree of convergence of nonlinear spline approximation*. In: Approximations with Special Emphasis on Spline Functions, ed. I.J. Schoenberg. Academic Press: New York, 349-365.

Richter G.R. (1976). *On weakly singular integral equations with displacement kernels*. J. Math. Anal. and Appl. 55, 32-42.

Richter G.R. (1978). *Superconvergence of piecewise polynomial Galerkin approximations for Fredholm equations of the second kind*. Numer. Math. 31, 63-70.

Saranen J. (1989). *Extrapolation methods for spline collocation solutions of pseudodifferentsial equations on curves*. Numer. Math. 56, 385-407.

Saranen J., Wendland W.L. (1985). *On the asymptotic convergence of collocation methods with spline functions of even degree*. Math. Comp. 45, 499-512.

Schatz A.H., Thomée V., Wendland W.L. (1990). *Mathematical Theory of Finite and Boundary Element Methods*. Birkhäuser: Basel-Boston-Berlin.

Schmitz H. (1989). *Über das singuläre Verhalten der Lösungen von Integralgleichungen auf Flächen mit Ecken*. Doctor thesis, Univ. Stuttgart.

Schneider C. (1979). *Regularity of the solution to a class of weakly singular Fredholm integral equations of the second kind*. Integral Equat. and Oper. Theory 2, 62-68.

Schneider C. (1981). *Product integration for weakly singular integral equations*. Math. Comp. 36, 207-213.

Sloan I. (1990). *Superconvergence*. In: Numerical Solution of Integral Equations, ed. M. Golberg, Plenum Press, 35-70.

Sloan I.H., Spence A. (1986). *Projection methods for integral equations on half-line*. IMA J. Numer. Anal., V.6, 153-172.

Sloan I.H., Thomée V. (1985). *Superconvergence of Galerkin iterates for integral equations of the second kind*. J. Integr. Eq. 9, 1-23.

Smelov V.V. (1978). *Lectures on the Neutron Transport*. Atomizdat: Moscow (in Russian).

Sobolev S.L. (1956). *Some remarks on the numerical solution of integral equations*. Izv. Acad. Sci. USSR, Ser Math. 20, No.4, 413-436 (in Russian).

Spence A., Thomas K.S. (1983). *On superconvergence properties of Galerkin's method for compact operator equations*. IMA J. Num. Anal. 3, 253-271.

Stummel F. (1970). *Diskrete Konvergenz linearer Operatoren, I*. Math. Ann. 190, No.1, 45-92.

Stummel F. (1971). *Diskrete Konvergenz linearer Operatoren, II*. Math. Z. 120, No.3, 231-264.

Stummel F. (1972). *Diskrete Konvergenz linearer Operatoren, III*. Proc. Oberwolfach Conf. in Linear Operators and Approximations. Birkhauser: Basel.

Stummel F. (1973). *Discrete convergence of mappings*. In: Top. Numer. Anal., London-New York, 285-310.

Stummel F. (1975). *Discretely uniform approximation of continuous functions*. J. Approxim. Theory 13, No.2, 178-191.

Stummel F. (1976). *Stable and discrete convergence of differentiable mapping*. Rev. roum. math. pures et appl. 21, No.1, 63-96.

Triebel H. (1978). *Interpolation Theory, Function Spaces, Differential Operators*. North-Holland: Amsterdam.

Uba P. (1983). *Iterative solution of an integral equation with weakly singular kernel*. Acta et comm. Univ. Tartuensis 633, 67-74 (in Russian).

Uba P. (1988a). *The smoothness of solution of weakly singular integral equations with a discontinuous coefficient*. Proc. Eston. Acad. Sci, Phys., Math. 37, No.2, 192-203 (in Russian).

Uba P. (1988b). *A numerical solution of a weakly singular integral equation with a discontinuous coefficient*. Acta et comm. Univ. Tartuensis 833, 28-34 (in Russian).

Uba P. (1989). *A collocation method with cubic splines to the solution of a multidimensional weakly singular integral equation*. Acta et comm. Univ. Tartuensis 863, 19-25.

Uba P. (1992). *Approximate computation of weakly singular integrals*. Acta et comm. Univ. Tartuensis 937, 27-35.

Vainikko G. (1968). *Compact approximation of linear compact operators in factor spaces*. Acta et comm. Univ. Tartuensis 220, 190-204 (in Russian).

Vainikko G. (1969). *The principle of compact approximation in the theory of approximate methods*. U.S.S.R. Comp. Math. and Math. Phys. 9, No 4, 739-761 (in Russian).

Vainikko G. (1970). *Compact Approximation of Operators and Approximate Solution of Equations*. University of Tartu: Tartu (in Russian).

Vainikko G. (1971). *On the convergence of quadrature formula method for integral equations with discontinuous kernels*. Siberian Math. J. 12, No.1, 40-53 (in Russian).

Vainikko G. (1974a). *On the approximation of fixed points of compact operators*. Acta et comm. Univ. Tartuensis 342, 225-236 (in Russian).

Vainikko G. (1974b). *Discretely compact sequences*. U.S.S.R. Comp. Math. and Math. Phys. 14, No.3, 572-583 (in Russian).

Vainikko G. (1976a). *Funktionalanalysis der Diskretisierungsmethoden*. Teubner: Leiptzig.

Vainikko G. (1976b). *Analysis of Discretization Methods*. Univ. of Tartu: Tartu (in Russian).

Vainikko G. (1977a). *Über die Konvergenz und Divergenz von Näherungsmethoden bei Eigenwertproblemen*. Math. Nachr. 78, 145-164.

Vainikko G. (1977b). *Über Konvergenzbegriffe für lineare Operatoren in der Numerischen Mathematik*. Math. Nachr. 78, 165-183.

Vainikko G. (1978). *Approximative methods for nonlinear equations*. Nonlin. Analysis 2, No.6, 647-687.

Vainikko G. (1981a). *Regular convergence of operators and approximate solution of equations*. J. Soviet Math. 15, No.6, 675-705. (Translation from Itogi Nauki, Ser. Math., Moscow, 1979, V.16, 5-53).

Vainikko G. (1981b). *Discrete measures of non-compactness*. Acta et comm. Univ. Tartuensis 580, 3-8 (in Russian).

Vainikko G. (1983). *On the invariance of the rotation of vector fields by the approximation of multivalued operators*. Acta et comm. Univ. Tartuensis 633, 3-10 (in Russian).

Vainikko G. (1988). *Piecewise polynomial approximation of a solution to multidimensional weakly singular integral equation*. Acta et comm. Univ. Tartuensis 833, 19-26 (in Russian).

Vainikko G. (1990a). *Collocation methods for multidimensional weakly singular integral equations*. In: Numer. Anal. and Math. Modelling. Banach center publ.: Warsaw, 91-105 (in Russian).

Vainikko G. (1990b). *Integral equations of an interior-exterior problem and approximate solution of them*. Proc. Eston. Acad. Sci., phys., math. 39, 185-195 (in Russian).

Vainikko G. (1990c). *Estimations of the derivatives of a solution to nonlinear weakly singular integral equation*. In: Problems of Pure and Applied Mathematics, Univ. of Tartu, 212-215.

Vainikko G. (1990d). *Radiation Transfer*. University of Tartu: Tartu (in Estonian).

Vainikko G. (1991a). *On the smoothness of the solution of multidimensional weakly singular integral equations*. Math. USSR Sbornik 68, No.2, 585-600. (Russian original 1989).

Vainikko G. (1991b). *Piecewise constant approximation of the solution to multidimensional weakly singular integral equations*. U.S.S.R. Comp. Math. and Math. Phys. 31, No.6, 832-849 (in Russian).

Vainikko G. (1992a). *Solution of large systems arising by discretization of multidimensional weakly singular integral equations*. Acta et comm. Univ. Tartuensis 937, 3-14.

Vainikko G. (1992b). *On the piecewise constant collocation method for multidimensional weakly singular integral equations.* J. Integral Eq. and Appl. 4, No.4.

Vainikko G., Karma O. (1974a). *On the convergence of approximate methods for linear and nonlinear operator equations.* U.S.S.R. Comp. Math. and Math. Phys. 14, No.4, 828-837 (in Russian).

Vainikko G., Karma O. (1974b). *On the convergence rate of approximate methods in eigenvalue problem with nonlinear dependence on the parameter.* U.S.S.R. Comp. Math. and Math. Phys. 14, No.6, 1393-1408 (in Russian).

Vainikko G., Karma O. (1988). An error estimation of approximate methods for *eigenvalue problems.* Acta et comm. Univ. Tartuensis 833, 75-83 (in Russian).

Vainikko G., Marshak A. (1978). *On the convergence rate of the discrete ordinate method in radiation transfer problem.* Soviet Math. (Iz. VUZ) No.11, 11-22 (in Russian).

Vainikko G., Pedas A. (1981). *The properties of solutions of weakly singular integral equations.* J. Austral. Math. Soc. (Ser. B) 22, 419-430.

Vainikko G., Pedas A. (1989). *A piecewise constant approximation to the solution of a weakly singular integral equation.* Acta et comm. Univ. Tartuensis 863, 31-39 (in Russian).

Vainikko G., Pedas A. (1990). *Convergence rate of a modified cubature formula method for multidimensional weakly singular integral equations.* Acta et comm. Univ. Tartuensis 913, 3-17.

Vainikko G., Pedas A., Uba P. (1984). *Methods for Solving Weakly Singular Integral Equations.* Univ. of Tartu: Tartu (in Russian).

Vainikko G., Piskarev S. (1977). *On regularly consistent operators.* Soviet Math. (Iz. VUZ) No.10, 25-36 (in Russian).

Vainikko G., Uba P. (1981). *A piecewise polynomial approximation to the solution of an integral equation with weakly singular kernel.* J. Austral. Math. Soc. (Ser. B) 22, 431-438.

Wendland W. (1980). *On Galerkin collocation methods for integral equations of elliptic boundary value problems.* In: Numerical Treatment of Integral Equations, ed. J. Albrecht and L. Collatz. Birkhäuser: Basel.

Wendland W. (1983). *Boundary element methods and their asymptotic convergence.* In: Theoretical Acoustics and Numerical Techniques, ed. P. Filippi. Springer-Verlag: Wien, New York, 135-216.

Wolf R. (1974). *Über lineare approximationsreguläre Operatoren.* Math. Nachr. 59, 325-341.

Wolf R. (1979). *Zur Stabilität der algebraischen Vielfachheit von Eigenwerten von holomorphen Fredholm-Operatorfunktionen.* Appl. Anal. 9, No.3, 165-177.

Yosida K. (1965). *Functional Analysis.* Springer-Verlag: Berlin, Göttingen, Heidelberg.

SUBJECT INDEX

Printing: Weihert-Druck GmbH, Darmstadt
Binding: Buchbinderei Schäffer, Grünstadt

Lecture Notes in Mathematics

For information about Vols. 1–1364
please contact your bookseller or Springer-Verlag

Vol. 1365: M. Giaquinta (Ed.), Topics in Calculus of Variations. Seminar, 1987. X, 196 pages. 1989.

Vol. 1366: N. Levitt, Grassmannians and Gauss Maps in PL-Topology. V, 203 pages. 1989.

Vol. 1367: M. Knebusch, Weakly Semialgebraic Spaces. XX, 376 pages. 1989.

Vol. 1368: R. Hübl, Traces of Differential Forms and Hochschild Homology. III, 111 pages. 1989.

Vol. 1369: B. Jiang, Ch.-K. Peng, Z. Hou (Eds.), Differential Geometry and Topology. Proceedings, 1986–87. VI, 366 pages. 1989.

Vol. 1370: G. Carlsson, R.L. Cohen, H.R. Miller, D.C. Ravenel (Eds.), Algebraic Topology. Proceedings, 1986. IX, 456 pages. 1989.

Vol. 1371: S. Glaz, Commutative Coherent Rings. XI, 347 pages. 1989.

Vol. 1372: J. Azéma, P.A. Meyer, M. Yor (Eds.), Séminaire de Probabilités XXIII. Proceedings. IV, 583 pages. 1989.

Vol. 1373: G. Benkart, J.M. Osborn (Eds.), Lie Algebras. Madison 1987. Proceedings. V, 145 pages. 1989.

Vol. 1374: R.C. Kirby, The Topology of 4-Manifolds. VI, 108 pages. 1989.

Vol. 1375: K. Kawakubo (Ed.), Transformation Groups. Proceedings, 1987. VIII, 394 pages, 1989.

Vol. 1376: J. Lindenstrauss, V.D. Milman (Eds.), Geometric Aspects of Functional Analysis. Seminar (GAFA) 1987–88. VII, 288 pages. 1989.

Vol. 1377: J.F. Pierce, Singularity Theory, Rod Theory, and Symmetry-Breaking Loads. IV, 177 pages. 1989.

Vol. 1378: R.S. Rumely, Capacity Theory on Algebraic Curves. III, 437 pages. 1989.

Vol. 1379: H. Heyer (Ed.), Probability Measures on Groups IX. Proceedings, 1988. VIII, 437 pages. 1989.

Vol. 1380: H.P. Schlickewei, E. Wirsing (Eds.), Number Theory, Ulm 1987. Proceedings. V, 266 pages. 1989.

Vol. 1381: J.-O. Strömberg, A. Torchinsky, Weighted Hardy Spaces. V, 193 pages. 1989.

Vol. 1382: H. Reiter, Metaplectic Groups and Segal Algebras. XI, 128 pages. 1989.

Vol. 1383: D.V. Chudnovsky, G.V. Chudnovsky, H. Cohn, M.B. Nathanson (Eds.), Number Theory, New York 1985–88. Seminar. V, 256 pages. 1989.

Vol. 1384: J. Garcia-Cuerva (Ed.), Harmonic Analysis and Partial Differential Equations. Proceedings, 1987. VII, 213 pages. 1989.

Vol. 1385: A.M. Anile, Y. Choquet-Bruhat (Eds.), Relativistic Fluid Dynamics. Seminar, 1987. V, 308 pages. 1989.

Vol. 1386: A. Bellen, C.W. Gear, E. Russo (Eds.), Numerical Methods for Ordinary Differential Equations. Proceedings, 1987. VII, 136 pages. 1989.

Vol. 1387: M. Petkovi´c, Iterative Methods for Simultaneous Inclusion of Polynomial Zeros. X, 263 pages. 1989.

Vol. 1388: J. Shinoda, T.A. Slaman, T. Tugué (Eds.), Mathematical Logic and Applications. Proceedings, 1987. V, 223 pages. 1989.

Vol. 1000: Second Edition. H. Hopf, Differential Geometry in the Large. VII, 184 pages. 1989.

Vol. 1389: E. Ballico, C. Ciliberto (Eds.), Algebraic Curves and Projective Geometry. Proceedings, 1988. V, 288 pages. 1989.

Vol. 1390: G. Da Prato, L. Tubaro (Eds.), Stochastic Partial Differential Equations and Applications II. Proceedings, 1988. VI, 258 pages. 1989.

Vol. 1391: S. Cambanis, A. Weron (Eds.), Probability Theory on Vector Spaces IV. Proceedings, 1987. VIII, 424 pages. 1989.

Vol. 1392: R. Silhol, Real Algebraic Surfaces. X, 215 pages. 1989.

Vol. 1393: N. Bouleau, D. Feyel, F. Hirsch, G. Mokobodzki (Eds.), Séminaire de Théorie du Potentiel Paris, No. 9. Proceedings. VI, 265 pages. 1989.

Vol. 1394: T.L. Gill, W.W. Zachary (Eds.), Nonlinear Semigroups, Partial Differential Equations and Attractors. Proceedings, 1987. IX, 233 pages. 1989.

Vol. 1395: K. Alladi (Ed.), Number Theory, Madras 1987. Proceedings. VII, 234 pages. 1989.

Vol. 1396: L. Accardi, W. von Waldenfels (Eds.), Quantum Probability and Applications IV. Proceedings, 1987. VI, 355 pages. 1989.

Vol. 1397: P.R. Turner (Ed.), Numerical Analysis and Parallel Processing. Seminar, 1987. VI, 264 pages. 1989.

Vol. 1398: A.C. Kim, B.H. Neumann (Eds.), Groups – Korea 1988. Proceedings. V, 189 pages. 1989.

Vol. 1399: W.-P. Barth, H. Lange (Eds.), Arithmetic of Complex Manifolds. Proceedings, 1988. V, 171 pages. 1989.

Vol. 1400: U. Jannsen. Mixed Motives and Algebraic K-Theory. XIII, 246 pages. 1990.

Vol. 1401: J. Steprans, S. Watson (Eds.), Set Theory and its Applications. Proceedings, 1987. V, 227 pages. 1989.

Vol. 1402: C. Carasso, P. Charrier, B. Hanouzet, J.-L. Joly (Eds.), Nonlinear Hyperbolic Problems. Proceedings, 1988. V, 249 pages. 1989.

Vol. 1403: B. Simeone (Ed.), Combinatorial Optimization. Seminar, 1986. V, 314 pages. 1989.

Vol. 1404: M.-P. Malliavin (Ed.), Séminaire d´Algèbre Paul Dubreil et Marie-Paul Malliavin. Proceedings, 1987–1988. IV, 410 pages. 1989.

Vol. 1405: S. Dolecki (Ed.), Optimization. Proceedings, 1988. V, 223 pages. 1989. Vol. 1406: L. Jacobsen (Ed.),

Analytic Theory of Continued Fractions III. Proceedings, 1988. VI, 142 pages. 1989.

Vol. 1407: W. Pohlers, Proof Theory. VI, 213 pages. 1989.

Vol. 1408: W. Lück, Transformation Groups and Algebraic K-Theory. XII, 443 pages. 1989.

Vol. 1409: E. Hairer, Ch. Lubich, M. Roche. The Numerical Solution of Differential-Algebraic Systems by Runge-Kutta Methods. VII, 139 pages. 1989.

Vol. 1410: F.J. Carreras, O. Gil-Medrano, A.M. Naveira (Eds.), Differential Geometry. Proceedings, 1988. V, 308 pages. 1989.

Vol. 1411: B. Jiang (Ed.), Topological Fixed Point Theory and Applications. Proceedings. 1988. VI, 203 pages. 1989.

Vol. 1412: V.V. Kalashnikov, V.M. Zolotarev (Eds.), Stability Problems for Stochastic Models. Proceedings, 1987. X, 380 pages. 1989.

Vol. 1413: S. Wright, Uniqueness of the Injective III_1 Factor. III, 108 pages. 1989.

Vol. 1414: E. Ramirez de Arellano (Ed.), Algebraic Geometry and Complex Analysis. Proceedings, 1987. VI, 180 pages. 1989.

Vol. 1415: M. Langevin, M. Waldschmidt (Eds.), Cinquante Ans de Polynômes. Fifty Years of Polynomials. Proceedings, 1988. IX, 235 pages.1990.

Vol. 1416: C. Albert (Ed.), Géométrie Symplectique et Mécanique. Proceedings, 1988. V, 289 pages. 1990.

Vol. 1417: A.J. Sommese, A. Biancofiore, E.L. Livorni (Eds.), Algebraic Geometry. Proceedings, 1988. V, 320 pages. 1990.

Vol. 1418: M. Mimura (Ed.), Homotopy Theory and Related Topics. Proceedings, 1988. V, 241 pages. 1990.

Vol. 1419: P.S. Bullen, P.Y. Lee, J.L. Mawhin, P. Muldowney, W.F. Pfeffer (Eds.), New Integrals. Proceedings, 1988. V, 202 pages. 1990.

Vol. 1420: M. Galbiati, A. Tognoli (Eds.), Real Analytic Geometry. Proceedings, 1988. IV, 366 pages. 1990.

Vol. 1421: H.A. Biagioni, A Nonlinear Theory of Generalized Functions, XII, 214 pages. 1990.

Vol. 1422: V. Villani (Ed.), Complex Geometry and Analysis. Proceedings, 1988. V, 109 pages. 1990.

Vol. 1423: S.O. Kochman, Stable Homotopy Groups of Spheres: A Computer-Assisted Approach. VIII, 330 pages. 1990.

Vol. 1424: F.E. Burstall, J.H. Rawnsley, Twistor Theory for Riemannian Symmetric Spaces. III, 112 pages. 1990.

Vol. 1425: R.A. Piccinini (Ed.), Groups of Self-Equivalences and Related Topics. Proceedings, 1988. V, 214 pages. 1990.

Vol. 1426: J. Azéma, P.A. Meyer, M. Yor (Eds.), Séminaire de Probabilités XXIV, 1988/89. V, 490 pages. 1990.

Vol. 1427: A. Ancona, D. Geman, N. Ikeda, École d'Eté de Probabilités de Saint Flour XVIII, 1988. Ed.: P.L. Hennequin. VII, 330 pages. 1990.

Vol. 1428: K. Erdmann, Blocks of Tame Representation Type and Related Algebras. XV. 312 pages. 1990.

Vol. 1429: S. Homer, A. Nerode, R.A. Platek, G.E. Sacks, A. Scedrov, Logic and Computer Science. Seminar, 1988. Editor: P. Odifreddi. V, 162 pages. 1990.

Vol. 1430: W. Bruns, A. Simis (Eds.), Commutative Algebra. Proceedings. 1988. V, 160 pages. 1990.

Vol. 1431: J.G. Heywood, K. Masuda, R. Rautmann, V.A. Solonnikov (Eds.), The Navier-Stokes Equations – Theory and Numerical Methods. Proceedings, 1988. VII, 238 pages. 1990.

Vol. 1432: K. Ambos-Spies, G.H. Müller, G.E. Sacks (Eds.), Recursion Theory Week. Proceedings, 1989. VI, 393 pages. 1990.

Vol. 1433: S. Lang, W. Cherry, Topics in Nevanlinna Theory. II, 174 pages.1990.

Vol. 1434: K. Nagasaka, E. Fouvry (Eds.), Analytic Number Theory. Proceedings, 1988. VI, 218 pages. 1990.

Vol. 1435: St. Ruscheweyh, E.B. Saff, L.C. Salinas, R.S. Varga (Eds.), Computational Methods and Function Theory. Proceedings, 1989. VI, 211 pages. 1990.

Vol. 1436: S. Xambó-Descamps (Ed.), Enumerative Geometry. Proceedings, 1987. V, 303 pages. 1990.

Vol. 1437: H. Inassaridze (Ed.), K-theory and Homological Algebra. Seminar, 1987–88. V, 313 pages. 1990.

Vol. 1438: P.G. Lemarié (Ed.) Les Ondelettes en 1989. Seminar. IV, 212 pages. 1990.

Vol. 1439: E. Bujalance, J.J. Etayo, J.M. Gamboa, G. Gromadzki. Automorphism Groups of Compact Bordered Klein Surfaces: A Combinatorial Approach. XIII, 201 pages. 1990.

Vol. 1440: P. Latiolais (Ed.), Topology and Combinatorial Groups Theory. Seminar, 1985–1988. VI, 207 pages. 1990.

Vol. 1441: M. Coornaert, T. Delzant, A. Papadopoulos. Géométrie et théorie des groupes. X, 165 pages. 1990.

Vol. 1442: L. Accardi, M. von Waldenfels (Eds.), Quantum Probability and Applications V. Proceedings, 1988. VI, 413 pages. 1990.

Vol. 1443: K.H. Dovermann, R. Schultz, Equivariant Surgery Theories and Their Periodicity Properties. VI, 227 pages. 1990.

Vol. 1444: H. Korezlioglu, A.S. Ustunel (Eds.), Stochastic Analysis and Related Topics VI. Proceedings, 1988. V, 268 pages. 1990.

Vol. 1445: F. Schulz, Regularity Theory for Quasilinear Elliptic Systems and – Monge Ampère Equations in Two Dimensions. XV, 123 pages. 1990.

Vol. 1446: Methods of Nonconvex Analysis. Seminar, 1989. Editor: A. Cellina. V, 206 pages. 1990.

Vol. 1447: J.-G. Labesse, J. Schwermer (Eds), Cohomology of Arithmetic Groups and Automorphic Forms. Proceedings, 1989. V, 358 pages. 1990.

Vol. 1448: S.K. Jain, S.R. López-Permouth (Eds.), Non-Commutative Ring Theory. Proceedings, 1989. V, 166 pages. 1990.

Vol. 1449: W. Odyniec, G. Lewicki, Minimal Projections in Banach Spaces. VIII, 168 pages. 1990.

Vol. 1450: H. Fujita, T. Ikebe, S.T. Kuroda (Eds.), Functional-Analytic Methods for Partial Differential Equations. Proceedings, 1989. VII, 252 pages. 1990.

Vol. 1451: L. Alvarez-Gaumé, E. Arbarello, C. De Concini, N.J. Hitchin, Global Geometry and Mathematical Physics. Montecatini Terme 1988. Seminar. Editors: M. Francaviglia, F. Gherardelli. IX, 197 pages. 1990.

Vol. 1452: E. Hlawka, R.F. Tichy (Eds.), Number-Theoretic Analysis. Seminar, 1988–89. V, 220 pages. 1990.

Vol. 1453: Yu.G. Borisovich, Yu.E. Gliklikh (Eds.), Global Analysis – Studies and Applications IV. V, 320 pages. 1990.

Vol. 1454: F. Baldassari, S. Bosch, B. Dwork (Eds.), p-adic Analysis. Proceedings, 1989. V, 382 pages. 1990.

Vol. 1455: J.-P. Françoise, R. Roussarie (Eds.), Bifurcations of Planar Vector Fields. Proceedings, 1989. VI, 396 pages. 1990.

Vol. 1456: L.G. Kovács (Ed.), Groups – Canberra 1989. Proceedings. XII, 198 pages. 1990.

Vol. 1457: O. Axelsson, L.Yu. Kolotilina (Eds.), Preconditioned Conjugate Gradient Methods. Proceedings, 1989. V, 196 pages. 1990.

Vol. 1458: R. Schaaf, Global Solution Branches of Two Point Boundary Value Problems. XIX, 141 pages. 1990.

Vol. 1459: D. Tiba, Optimal Control of Nonsmooth Distributed Parameter Systems. VII, 159 pages. 1990.

Vol. 1460: G. Toscani, V. Boffi, S. Rionero (Eds.), Mathematical Aspects of Fluid Plasma Dynamics. Proceedings, 1988. V, 221 pages. 1991.

Vol. 1461: R. Gorenflo, S. Vessella, Abel Integral Equations. VII, 215 pages. 1991.

Vol. 1462: D. Mond, J. Montaldi (Eds.), Singularity Theory and its Applications. Warwick 1989, Part I. VIII, 405 pages. 1991.

Vol. 1463: R. Roberts, I. Stewart (Eds.), Singularity Theory and its Applications. Warwick 1989, Part II. VIII, 322 pages. 1991.

Vol. 1464: D. L. Burkholder, E. Pardoux, A. Sznitman, Ecole d'Eté de Probabilités de Saint- Flour XIX-1989. Editor: P. L. Hennequin. VI, 256 pages. 1991.

Vol. 1465: G. David, Wavelets and Singular Integrals on Curves and Surfaces. X, 107 pages. 1991.

Vol. 1466: W. Banaszczyk, Additive Subgroups of Topological Vector Spaces. VII, 178 pages. 1991.

Vol. 1467: W. M. Schmidt, Diophantine Approximations and Diophantine Equations. VIII, 217 pages. 1991.

Vol. 1468: J. Noguchi, T. Ohsawa (Eds.), Prospects in Complex Geometry. Proceedings, 1989. VII, 421 pages. 1991.

Vol. 1469: J. Lindenstrauss, V. D. Milman (Eds.), Geometric Aspects of Functional Analysis. Seminar 1989-90. XI, 191 pages. 1991.

Vol. 1470: E. Odell, H. Rosenthal (Eds.), Functional Analysis. Proceedings, 1987-89. VII, 199 pages. 1991.

Vol. 1471: A. A. Panchishkin, Non-Archimedean L-Functions of Siegel and Hilbert Modular Forms. VII, 157 pages. 1991.

Vol. 1472: T. T. Nielsen, Bose Algebras: The Complex and Real Wave Representations. V, 132 pages. 1991.

Vol. 1473: Y. Hino, S. Murakami, T. Naito, Functional Differential Equations with Infinite Delay. X, 317 pages. 1991.

Vol. 1474: S. Jackowski, B. Oliver, K. Pawałowski (Eds.), Algebraic Topology, Poznań 1989. Proceedings. VIII, 397 pages. 1991.

Vol. 1475: S. Busenberg, M. Martelli (Eds.), Delay Differential Equations and Dynamical Systems. Proceedings, 1990. VIII, 249 pages. 1991.

Vol. 1476: M. Bekkali, Topics in Set Theory. VII, 120 pages. 1991.

Vol. 1477: R. Jajte, Strong Limit Theorems in Noncommutative L_2-Spaces. X, 113 pages. 1991.

Vol. 1478: M.-P. Malliavin (Ed.), Topics in Invariant Theory. Seminar 1989-1990. VI, 272 pages. 1991.

Vol. 1479: S. Bloch, I. Dolgachev, W. Fulton (Eds.), Algebraic Geometry. Proceedings, 1989. VII, 300 pages. 1991.

Vol. 1480: F. Dumortier, R. Roussarie, J. Sotomayor, H. Żoładek, Bifurcations of Planar Vector Fields: Nilpotent Singularities and Abelian Integrals. VIII, 226 pages. 1991.

Vol. 1481: D. Ferus, U. Pinkall, U. Simon, B. Wegner (Eds.), Global Differential Geometry and Global Analysis. Proceedings, 1991. VIII, 283 pages. 1991.

Vol. 1482: J. Chabrowski, The Dirichlet Problem with L^2-Boundary Data for Elliptic Linear Equations. VI, 173 pages. 1991.

Vol. 1483: E. Reithmeier, Periodic Solutions of Nonlinear Dynamical Systems. VI, 171 pages. 1991.

Vol. 1484: H. Delfs, Homology of Locally Semialgebraic Spaces. IX, 136 pages. 1991.

Vol. 1485: J. Azéma, P. A. Meyer, M. Yor (Eds.), Séminaire de Probabilités XXV. VIII, 440 pages. 1991.

Vol. 1486: L. Arnold, H. Crauel, J.-P. Eckmann (Eds.), Lyapunov Exponents. Proceedings, 1990. VIII, 365 pages. 1991.

Vol. 1487: E. Freitag, Singular Modular Forms and Theta Relations. VI, 172 pages. 1991.

Vol. 1488: A. Carboni, M. C. Pedicchio, G. Rosolini (Eds.), Category Theory. Proceedings, 1990. VII, 494 pages. 1991.

Vol. 1489: A. Mielke, Hamiltonian and Lagrangian Flows on Center Manifolds. X, 140 pages. 1991.

Vol. 1490: K. Metsch, Linear Spaces with Few Lines. XIII, 196 pages. 1991.

Vol. 1491: E. Lluis-Puebla, J.-L. Loday, H. Gillet, C. Soulé, V. Snaith, Higher Algebraic K-Theory: an overview. IX, 164 pages. 1992.

Vol. 1492: K. R. Wicks, Fractals and Hyperspaces. VIII, 168 pages. 1991.

Vol. 1493: E. Benoît (Ed.), Dynamic Bifurcations. Proceedings, Luminy 1990. VII, 219 pages. 1991.

Vol. 1494: M.-T. Cheng, X.-W. Zhou, D.-G. Deng (Eds.), Harmonic Analysis. Proceedings, 1988. IX, 226 pages. 1991.

Vol. 1495: J. M. Bony, G. Grubb, L. Hörmander, H. Komatsu, J. Sjöstrand, Microlocal Analysis and Applications. Montecatini Terme, 1989. Editors: L. Cattabriga, L. Rodino. VII, 349 pages. 1991.

Vol. 1496: C. Foias, B. Francis, J. W. Helton, H. Kwakernaak, J. B. Pearson, H_∞-Control Theory. Como, 1990. Editors: E. Mosca, L. Pandolfi. VII, 336 pages. 1991.

Vol. 1497: G. T. Herman, A. K. Louis, F. Natterer (Eds.), Mathematical Methods in Tomography. Proceedings 1990. X, 268 pages. 1991.

Vol. 1498: R. Lang, Spectral Theory of Random Schrödinger Operators. X, 125 pages. 1991.

Vol. 1499: K. Taira, Boundary Value Problems and Markov Processes. IX, 132 pages. 1991.

Vol. 1500: J.-P. Serre, Lie Algebras and Lie Groups. VII, 168 pages. 1992.

Vol. 1501: A. De Masi, E. Presutti, Mathematical Methods for Hydrodynamic Limits. IX, 196 pages. 1991.

Vol. 1502: C. Simpson, Asymptotic Behavior of Monodromy. V, 139 pages. 1991.

Vol. 1503: S. Shokranian, The Selberg-Arthur Trace Formula (Lectures by J. Arthur). VII, 97 pages. 1991.

Vol. 1504: J. Cheeger, M. Gromov, C. Okonek, P. Pansu, Geometric Topology: Recent Developments. Editors: P. de Bartolomeis, F. Tricerri. VII, 197 pages. 1991.

Vol. 1505: K. Kajitani, T. Nishitani, The Hyperbolic Cauchy Problem. VII, 168 pages. 1991.

Vol. 1506: A. Buium, Differential Algebraic Groups of Finite Dimension. XV, 145 pages. 1992.

Vol. 1507: K. Hulek, T. Peternell, M. Schneider, F.-O. Schreyer (Eds.), Complex Algebraic Varieties. Proceedings, 1990. VII, 179 pages. 1992.

Vol. 1508: M. Vuorinen (Ed.), Quasiconformal Space Mappings. A Collection of Surveys 1960-1990. IX, 148 pages. 1992.

Vol. 1509: J. Aguadé, M. Castellet, F. R. Cohen (Eds.), Algebraic Topology - Homotopy and Group Cohomology. Proceedings, 1990. X, 330 pages. 1992.

Vol. 1510: P. P. Kulish (Ed.), Quantum Groups. Proceedings, 1990. XII, 398 pages. 1992.

Vol. 1511: B. S. Yadav, D. Singh (Eds.), Functional Analysis and Operator Theory. Proceedings, 1990. VIII, 223 pages. 1992.

Vol. 1512: L. M. Adleman, M.-D. A. Huang, Primality Testing and Abelian Varieties Over Finite Fields. VII, 142 pages. 1992.

Vol. 1513: L. S. Block, W. A. Coppel, Dynamics in One Dimension. VIII, 249 pages. 1992.

Vol. 1514: U. Krengel, K. Richter, V. Warstat (Eds.), Ergodic Theory and Related Topics III, Proceedings, 1990. VIII, 236 pages. 1992.

Vol. 1515: E. Ballico, F. Catanese, C. Ciliberto (Eds.), Classification of Irregular Varieties. Proceedings, 1990. VII, 149 pages. 1992.

Vol. 1516: R. A. Lorentz, Multivariate Birkhoff Interpolation. IX, 192 pages. 1992.

Vol. 1517: K. Keimel, W. Roth, Ordered Cones and Approximation. VI, 134 pages. 1992.

Vol. 1518: H. Stichtenoth, M. A. Tsfasman (Eds.), Coding Theory and Algebraic Geometry. Proceedings, 1991. VIII, 223 pages. 1992.

Vol. 1519: M. W. Short, The Primitive Soluble Permutation Groups of Degree less than 256. IX, 145 pages. 1992.

Vol. 1520: Yu. G. Borisovich, Yu. E. Gliklikh (Eds.), Global Analysis – Studies and Applications V. VII, 284 pages. 1992.

Vol. 1521: S. Busenberg, B. Forte, H. K. Kuiken, Mathematical Modelling of Industrial Process. Bari, 1990. Editors: V. Capasso, A. Fasano. VII, 162 pages. 1992.

Vol. 1522: J.-M. Delort, F. B. I. Transformation. VII, 101 pages. 1992.

Vol. 1523: W. Xue, Rings with Morita Duality. X, 168 pages. 1992.

Vol. 1524: M. Coste, L. Mahé, M.-F. Roy (Eds.), Real Algebraic Geometry. Proceedings, 1991. VIII, 418 pages. 1992.

Vol. 1525: C. Casacuberta, M. Castellet (Eds.), Mathematical Research Today and Tomorrow. VII, 112 pages. 1992.

Vol. 1526: J. Azéma, P. A. Meyer, M. Yor (Eds.), Séminaire de Probabilités XXVI. X, 633 pages. 1992.

Vol. 1527: M. I. Freidlin, J.-F. Le Gall, Ecole d'Eté de Probabilités de Saint-Flour XX – 1990. Editor: P. L. Hennequin. VIII, 244 pages. 1992.

Vol. 1528: G. Isac, Complementarity Problems. VI, 297 pages. 1992.

Vol. 1529: J. van Neerven, The Adjoint of a Semigroup of Linear Operators. X, 195 pages. 1992.

Vol. 1530: J. G. Heywood, K. Masuda, R. Rautmann, S. A. Solonnikov (Eds.), The Navier-Stokes Equations II – Theory and Numerical Methods. IX, 322 pages. 1992.

Vol. 1531: M. Stoer, Design of Survivable Networks. IV, 206 pages. 1992.

Vol. 1532: J. F. Colombeau, Multiplication of Distributions. X, 184 pages. 1992.

Vol. 1533: P. Jipsen, H. Rose, Varieties of Lattices. X, 162 pages. 1992.

Vol. 1534: C. Greither, Cyclic Galois Extensions of Commutative Rings. X, 145 pages. 1992.

Vol. 1535: A. B. Evans, Orthomorphism Graphs of Groups. VIII, 114 pages. 1992.

Vol. 1536: M. K. Kwong, A. Zettl, Norm Inequalities for Derivatives and Differences. VII, 150 pages. 1992.

Vol. 1537: P. Fitzpatrick, M. Martelli, J. Mawhin, R. Nussbaum, Topological Methods for Ordinary Differential Equations. Montecatini Terme, 1991. Editors: M. Furi, P. Zecca. VII, 218 pages. 1993.

Vol. 1538: P.-A. Meyer, Quantum Probability for Probabilists. X, 287 pages. 1993.

Vol. 1539: M. Coornaert, A. Papadopoulos, Symbolic Dynamics and Hyperbolic Groups. VIII, 138 pages. 1993.

Vol. 1540: H. Komatsu (Ed.), Functional Analysis and Related Topics, 1991. Proceedings. XXI, 413 pages. 1993.

Vol. 1541: D. A. Dawson, B. Maisonneuve, J. Spencer, Ecole d' Eté de Probabilités de Saint-Flour XXI - 1991. Editor: P. L. Hennequin. VIII, 356 pages. 1993.

Vol. 1542: J.Fröhlich, Th.Kerler, Quantum Groups, Quantum Categories and Quantum Field Theory. VII, 431 pages. 1993.

Vol. 1543: A. L. Dontchev, T. Zolezzi, Well-Posed Optimization Problems. XII, 421 pages. 1993.

Vol. 1544: M.Schürmann, White Noise on Bialgebras. VII, 146 pages. 1993.

Vol. 1545: J. Morgan, K. O'Grady, Differential Topology of Complex Surfaces. VIII, 224 pages. 1993.

Vol. 1546: V. V. Kalashnikov, V. M. Zolotarev (Eds.), Stability Problems for Stochastic Models. Proceedings, 1991. VIII, 229 pages. 1993.

Vol. 1547: P. Harmand, D. Werner, W. Werner, M-ideals in Banach Spaces and Banach Algebras. VIII, 387 pages. 1993.

Vol. 1548: T. Urabe, Dynkin Graphs and Quadrilateral Singularities. VI, 233 pages. 1993.

Vol. 1549: G. Vainikko, Multidimensional Weakly Singular Integral Equations. XI, 159 pages. 1993.